WHY ECOLOGY MATTERS

CHARLES J. KREBS

Why Ecology Matters

THE UNIVERSITY

OF CHICAGO PRESS

Chicago and London

Charles J. Krebs is professor emeritus of zoology
at the University of British Columbia and thinker
in residence in the Institute for Applied Ecology
at the University of Canberra.

The University of Chicago Press, Chicago 60637
The University of Chicago Press, Ltd., London
© 2016 by The University of Chicago
All rights reserved. Published 2016.
Printed in the United States of America

25 24 23 22 21 20 19 18 17 16   1 2 3 4 5

ISBN-13: 978-0-226-31801-1 (cloth)
ISBN-13: 978-0-226-31815-8 (paper)
ISBN-13: 978-0-226-31829-5 (e-book)
DOI: 10.7208/chicago/9780226318295.001.0001

Library of Congress Cataloging-in-Publication Data
Krebs, Charles J., author.
  Why ecology matters / Charles J. Krebs.
    pages cm
  Includes bibliographical references and index.
    ISBN 978-0-226-31801-1 (cloth : alk. paper) —
ISBN 978-0-226-31815-8 (pbk. : alk. paper) —
ISBN 978-0-226-31829-5 (e-book) 1. Ecology.
2. Population biology. 3. Coexistence of species.
I. Title.
  QH541.K677 2016
  577—dc23

                          2015033563

♾ This paper meets the requirements of
ANSI/NISO Z39.48-1992 (Permanence of Paper)

CONTENTS

PREFACE

Many of the discussions of the day involve the subject of ecology, and for that reason we have a vital interest in knowing some its principles. Everyone needs to be an ecologist to understand the biological issues that arise over climate change, disease outbreaks, endangered species, and the origins of antibiotic resistance in bacteria. Ecology is the science that deals with the general question of how all the animals and plants of Earth interact with their environment; we have serious problems when we ignore its basic principles. If we eliminate wolves, we may have an excess of deer to manage. If we have an excess of deer we may have an excess of ticks carrying Lyme disease, and humans may begin to get ill. If we use antibiotics on farmed chickens we may select for pathogenic bacteria that are resistant to all known antibiotics. Evolutionary changes can occur rapidly, and when they do so they have ecological consequences.

We all learn when we are young that we cannot ignore the basic principles of physics—we can walk and run but we cannot fly like the birds. We rely then on technology to produce airplanes to overcome this limitation. But what kinds of principles exist in ecology, and what do scientists know about them? That is the subject of this book. How does the natural world work? Is everything connected to everything else in nature? No, an ecologist would answer, but some species are connected to others, and by describing and understanding these connections we can manage our impacts on the Earth much better. If you aspire to be a dentist or an accountant or an astronaut you will still be tied to ecological systems about which some knowledge is useful. No matter what we do, we must eat food produced in agricultural enterprises that are a form of applied ecology, we must breathe the air that we have and realize that it is purified by plants, we must live in a healthy climate and should be concerned about the rising carbon dioxide level from fossil fuels. We may not be directly involved in protecting biodiversity in the tropics, but we might be sympathetic to the search for plant chemicals that can alleviate cancer or other emerging diseases.

There are many books on ecology and sources on the Internet that you could read. The best ones are highly specialized on the details of a few species of birds or butterflies, and these are most useful and readable.

Textbooks on ecology are usually very sound in details that are needed by a professional ecologist but too much for someone who is not a specialist. The Internet is useful for some particular information, but it does not offer an overview of ecological science. This book tries to hit the middle ground by providing an overview sprinkled with details. I have gathered 12 principles that cover the main ideas that are important in understanding how the biological world works. I have avoided jargon as much as possible, and tried to give examples of ecological insights that are presented in enough detail to capture the key ideas and issues that are unresolved at the present time. Ecology could be classified as a slow science in that it depends on data gathered over long time periods. Experiments that we hear about in the media are often carried out in a few months or years. Ecological insight may come only after 10–30 years of detailed studies and experiments. Glacial geologists used to laugh when they would discuss who could sit and watch a glacier move, but in these days of rapid climate change this is no longer a joke. Ecological science, in a similar way, is moving more rapidly as more and more research on how the world works is carried out around the globe.

I have arranged the main ideas of ecology into 12 chapters that proceed from a simple question of the geographical distribution of life on Earth to problems of populations, and then to more difficult issues of how groups of organisms get along in communities and ecosystems. So complexity grows as you move through these chapters. Much is known but much more is not yet well understood. In each chapter of this book I provide references to the scientific literature in case a particular question captures your interest and deserves follow-up. The ecological literature grows larger every day, and the insights that ecologists unravel enrich our lives by increasing our understanding of what is going on outside the windows of our house.

I am indebted to Alice Kenney for her assistance in helping to gather the material. There are two sets of heroes and heroines in ecology. The first set is all the field ecologists who have worked long hours to unravel the secrets of the ecosystems of the Earth. They are little appreciated, they get no Nobel Prizes, yet they are the scientists that our grandchildren will most appreciate. The second set of heroes and heroines are those providing funding for ecological research—governmental agencies that have made the brave decision to fund ecological research, knowing that some of the findings will be embarrassing to current government policies—and the private individuals and foundations who think ecological research well worth supporting. Little ecological research generates dollar bills, so it is often overlooked as a science of little importance in a world run by the mania of economic growth. But it enriches our lives, and perhaps that is more important than money.

CHARLES J. KREBS

CHAPTER 1

WHAT LIMITS THE GEOGRAPHIC DISTRIBUTION OF ORGANISMS?

KEY POINTS

- The distributions of many species are limited by geography and climate. In the past it was difficult for most species to move between continents. But humans are now moving species into new regions where some become serious pests.
- Climatic warming is also changing the distributions of many species, causing many ranges to expand toward the poles.
- On a very local level what limits the exact geographical ranges of species is not always clearly understood, and many ecological processes may be involved.

Penguins occur neither in Chicago nor in the Arctic. We are not particularly surprised about their absence in Chicago. Penguins hunt off the Antarctic ice pack and in the Southern Ocean for fish, and Chicago has neither pack ice nor an ocean. But penguins live happily in the Chicago zoo, so clearly the climate of Chicago is not the restricting factor. We should be surprised that penguins do not live in the Arctic, since it abounds with both ice packs and small fish, yet the reason is simple. Penguins have never reached the Arctic because the tropical oceans form a *barrier* that they have not been able to cross to enter the Arctic Ocean.

Barriers prevent dispersal movements, in particular the movement of an individual from its place of birth to a new place for breeding and reproduction. Movement is crucial in many ecological situations, but nowhere are the effects of movements more clearly shown than in the study of distribution. Isolation, or lack of dispersal, thus became a cornerstone of the early naturalists' view of how the animals and plants of the world came to be. This isolation is thus the reason we go to Africa to see giraffes and not to South America, and why we go to Australia to see kangaroos and not to North America. Our zoos are thus a popular monument to the role of dispersal in affecting the distribution of animal life on the globe just as our botanical gardens illustrate the same ideas about the distribution of plant life.

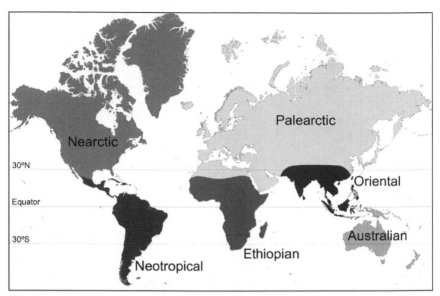

Figure 1.1 The Earth's biogeographic realms. These six broad regions are a product of continental drift over the last 200 million years and of barriers such as mountain ranges that have affected evolutionary processes. They were first recognized by Alfred Wallace (1876) and have been updated by Holt et al. (2013).

Alfred Wallace in 1876 outlined the broad pattern of the distribution of species on Earth with a classic view of the globe, divided into regions based mainly on the mammal fauna. Wallace distinguished North America (Nearctic) from Eurasia (Palearctic), and defined four other regions that divided the mammal fauna of the globe—South America (Neotropical), Africa (Ethiopian), Australia, and the Indian Subcontinent (Oriental) (Figure 1.1). Wallace recognized the patterns we see today when we go to Africa to see giraffes and to Australia to see kangaroos. This global view of the distribution of life has been the basis of the analysis of geographical distributions of animals, plants, and microbes, and provides a good starting point for understanding species ranges. It is a pattern written by the isolation of continents and regions by geographic barriers, leading to different evolutionary paths and thus different assemblages of species. It is the starting point for trying to understand why a particular species lives in a specific region, and also for understanding what the consequences might be of moving species across these boundaries.

But a problem arises here. Evolution has certainly produced different plants and animals in different geographical realms, but what assurance

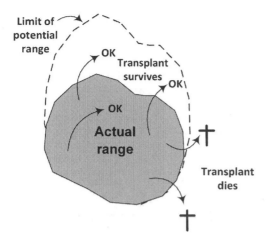

Figure 1.2 Hypothetical sets of transplant experiments. The grey area represents the actual current geographical range of a particular species. Each arrow indicates a transplant experiment. Arrows indicate successful transplants (OK) or unsuccessful transplants (†). In this example the species can potentially occupy a larger range (enclosed within the dashed line) than it currently does. In practice, many separate transplant experiments may be needed to define the limits of a species' potential geographic range.

do we have that any one of these organisms could in fact live in a quite different area? This question can be answered very simply by a transplant experiment—move the organism to a new area. If it survives there and reproduces, you have good evidence that the former distribution was restricted by a lack of dispersal. Figure 1.2 illustrates the logic of the simple transplant experiment.

People have carried out transplant experiments, often inadvertently, since the earliest times, but in the last two centuries this trickle of transfers has turned into a flood. Most of the crops we grow are introduced species of plants, and so transplant experiments can benefit humans. But many of our serious pests are also introduced species, and the ecology of invasive species has a strong economic impact on our lives. Many of the pest species transplanted are accidentals—seeds caught in bales of wool, or mice transported in bales of hay. An elaborate series of inspection and quarantine procedures in different nations illustrates how people strive to prevent the accidental or deliberate introduction of organisms harmful to humans and their domestic animals from one region to another.

Paradoxically, some of the worst pest species have been introduced deliberately. Consider just two examples. The European starling (*Sturnus vulgaris*) has spread over the entire United States and much of Canada within a period of sixty years. The starling is considered a pest because it is bold and aggressive, attacks some fruit crops, and has displaced several native bird species. Originally it occurred in Eurasia, from the Mediterranean to Norway and east to Siberia. Many early attempts were made to introduce the starling into the United States. One attempt was made at

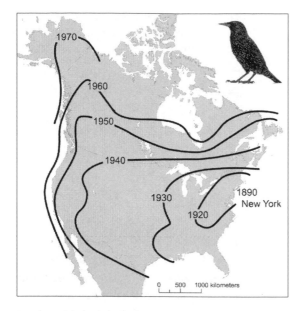

Figure 1.3 Westward expansion of the geographic range of the European starling (*Sturnus vulgaris*) in North America. The starling was introduced into New York City in 1890 and spread rapidly westward and northward. (Modified from Johnson and Cowan 1974.)

West Chester, Pennsylvania, before 1850 and the next at Cincinnati, Ohio, in 1872–1873, but nothing came of these or several other importations. In 1889 twenty pairs were released in Portland, Oregon, but these gradually disappeared. No one knows why these early introductions failed—perhaps too few individuals were released.

The permanent establishment of the starling in the United States dates from April 1890, when eighty birds were released in Central Park, New York City, by the president of the American Acclimatization Society, which tried to introduce every bird species mentioned in the works of William Shakespeare into North America. In March of the following year eighty more were released. About ten years were required for the starling to become established in the New York City area. It has since expanded its range across North America (Figure 1.3). This rapid expansion of the breeding range has been due to the irregular migrations and wanderings of nonbreeding juvenile birds, one and two years of age. Adult starlings typically use the same breeding area from year to year and thus do not colonize new areas. About three million square miles were colonized by the starling during the first fifty years after its successful introduction, and a bird unknown to our forefathers has now become one of the more common birds in North America.

The cane toad (*Rhinella marinus*) is native to Central and South America from Mexico to Brazil. It was widely introduced during the 1930s to islands in the Caribbean and the Pacific because it was believed to control scarab beetles, an insect pest of sugarcane. It was brought into northeast Queensland, Australia, in 1935, where it failed to control any insect pests and instead became a pest itself. Cane toads have parotid glands that contain a poison that causes cardiac arrest. All forms of the toad are poisonous, and humans eating cane toad eggs have died from the toxin. Cane toads eat almost anything but mainly insects, often those insects that do more good than harm. What they do not do is control the insect pests of sugarcane, the original justification for their introduction. They breed prolifically, females laying 8,000–35,000 eggs at least twice a year

Cane toads are toxic to many of their potential predators, but some species learn to avoid eating them or evolve resistance to the toxin. Because of their toxicity and high reproductive rate, cane toads have been moving across northern Australia since their introduction in 1935 (Figure 1.4). Cane toads have been moving west at about 40 kilometers per year and in 2009 crossed into Western Australia. Individual marked toads have moved up to 1.8 kilometers per night, primarily along roads that have served as convenient habitat corridors for rapid spread.

Cane toads must breed in small ponds, and one way to halt their spread into much of western Australia is to eliminate water holes in critical areas. Tingley et al. (2013) identified three points along the coastline of northwestern Western Australia that could be critical barriers to the spread of cane toads further south. Eliminating artificial water bodies in these areas would be highly effective in stopping the continued expansion of the range of the cane toad in Australia. The problem is that most of the water bodies that would have to be drained are on pastoral lands and are thus unlikely to be implemented because of economic losses to the immediate landholder.

Since cane toads are toxic in all their life history stages from eggs to tadpoles to toads, there was considerable worry during the 1990s and 2000s that their invasion shown in Figure 1.4 would cause massive mortality to predatory birds, reptiles, and mammals. Fortunately the impact of this toxic pest has not been as severe as was anticipated (Shine 2010). Populations of large predators such as lizards, elapid snakes, and freshwater crocodiles have been reduced temporarily by the cane toad invasion, but poisoning impacts are highly variable. Some of the predators severely reduced by toad invasion (like freshwater crocodiles) have recovered within a few decades, via learning to avoid eating cane toads. No native predators have gone extinct as a result of toad invasion, and many native taxa widely

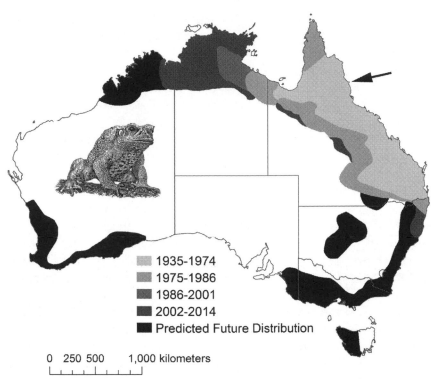

Figure 1.4 The spread of the introduced cane toad (*Rhinella marinus*) from its introduction in 1935 in Queensland (arrow) to 2014 and its predicted future spread to suitable areas in Southern and Western Australia. (After Urban et al. 2007 and data from Western Australia Parks and Wildlife Department 2014.)

imagined to be at risk are not affected, largely as a result of their physiological ability to tolerate toad toxins, as well as the reluctance of many native amphibian-eating predators to consume cane toads, either innately or as a learned response. The general conclusion of a modest impact by this introduced pest has to be tempered by the fact that detailed data on the populations of its predators and competitors, as well as the insects eaten by cane toads, was largely lacking. Ideally ecologists need before-and-after data to evaluate the impact of any introduced species, and little of this has been available for most pest species.

The other message left by the cane toad has been the warning that we should not introduce species in the belief that they are beneficial without very extensive study. Too many "desirable" introductions over the last two centuries have turned out to be ecological disasters.

Not all introduction experiments have harmful results, and one of the challenges of ecology is to sort out the positive and the negative before the transplant is done. We benefit from many introduced species—most of our agricultural crops qualify as successful transplant experiments. Many fishes have been introduced into new areas successfully, with a resulting improvement in fishing. The rainbow trout (*Oncorhynchus mykiss*) is a native of cool rivers and streams of western North America, and a prize game fish among fishermen. Rainbow trout have been introduced all over the globe during the last hundred years, and are now firmly established on all continents except Antarctica. Although originally the rainbow trout did not occur east of the Continental Divide in North America, it now occupies streams in all the Canadian provinces and most of the United States, as well as some of the river systems in Mexico and Central America. Trout fishing has expanded greatly because of these introductions. But even these apparently desirable introductions may have undesirable side effects in some regions. For example, rainbow trout can displace native brook trout, another prized game fish, in the southern Appalachians.

Not all transplant experiments are successful, and the dramatic effects of the successful transplants, such as the starling in North America, tend to overshadow the humdrum failures of many other introductions. Considerable historical research has been done on introductions of birds and mammals into Australia and New Zealand by acclimatization societies whose major purpose was to make New Zealand and Australia more like Europe and North America.

Many exotic species of birds and mammals were introduced into New Zealand during the 19th century. Acclimatization societies in some areas kept meticulous records of how many birds of each species were brought in and released in each year. One of the many findings from these careful records has been the observation that if more individuals of a species were introduced, the species was more likely to survive and colonize the island (Figure 1.5). This finding has become a cornerstone of a set of generalizations about invasive species introductions—more releases increase the likelihood of success. Small populations face a variety of chance events that can lead to extinction—bad weather or predator attacks that kill only a few individuals but tip the balance toward failure. Of 133 exotic bird species brought to New Zealand only about 45% survived to become permanent residents.

But as with many generalizations in ecology, there are exceptions. Sambar deer (*Cervus unicolor*) were introduced into New Zealand successfully

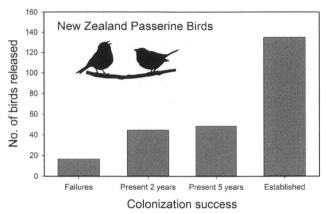

Figure 1.5 Average number of passerine birds released in New Zealand during the 19th century in relation to colonization success. The greater the numbers of individuals released, the more likely the success for any particular species. (Data from Blackburn et al. 2013.)

with only 2 individuals, and Himalayan tahr (*Hemitragus jemlahicus*) was successful with an introduction of 5 individuals in 1904. But in general for ungulates with adequate data, 11 of 14 species established and as we have just seen for birds, the more individuals released in general the higher the success rate of colonization.

The problems of invasive species have highlighted the general processes by which species can move into new areas. These processes are complicated and this is the reason why we have at present few general explanations about success or failure of introduced species. There are four major steps of the invasion process—transport, establishment, spread, and impact—and the invasion process can fail at any of these four steps. The final impact of the invasive species may be large or small, and the impact depends in part on human perception.

Transplants or movements of plants and animals into a new area may fail for two general reasons: either the biological environment may eliminate the newcomer or the physical-chemical environment may be lethal to the organism or prevent it from reproducing. Predators may prevent the establishment of some species. A good illustration of the role of predators can be seen in the common mussel (*Mytilus edulis*), which lives attached to rocks along sea coasts throughout the world. On the exposed southern coast of Ireland small mussels are abundant, but in protected waters mussels are often absent. The reason for this can be seen very easily if one moves pieces of rock with mussels attached from exposed coast to protected waters. Mussels disappear rapidly from protected waters because they are eaten by three species of crabs and a starfish. If you transplant the mussels to protected waters and put them inside a wire mesh cage, they will live happily as long as the predators cannot get into the cage. The crabs and the starfish are uncommon on the open coast because of heavy wave action in the intertidal zone, and the mussels thus have a refuge where they are relatively safe.

The expansion or contraction of geographical ranges is an important topic now because of climate change. Increases in carbon dioxide and other greenhouse gases in the atmosphere have triggered a gradual warming of the climate and changes in the distribution of rainfall. Global warming has left a strong imprint on the geographical ranges of many species (Burrows et al. 2014, Cahill et al. 2014). A combined analysis of 1,367 species responses around the world produced an average movement away from the equator of 18 kilometers per decade. This analysis covered plants, mammals, birds, beetles, grasshoppers, butterflies, intertidal algae and invertebrates, and spiders, and the average length of observations was 25 years (Chen et al. 2011). Similar data for movements higher up mountains averaged 12 m elevation per decade, and the average length of observations was 35 years. Detailed data on range boundary changes in spiders and butterflies from Britain are shown in Figure 1.6.

If climatic factors are the only explanation for changes in geographic distributions, we would expect all species to shift as climate warms. This is not the case because a whole range of factors can affect range limits. Changes in distributions for any particular species could be due to many ecological processes:

- Is the species absent because it has not been able to move to an area (dispersal limitation)?
- Is the species absent because it does not recognize the habitat as suitable?
- Do other species prevent colonization (parasites, predators, pathogens)?
- Are there limiting physical or chemical factors (temperature, water, oxygen, soil, pH)?

Changes in distribution because of climatic warming can be accepted only if the first three questions are carefully considered.

Large-scale patterns can obscure some of the observed shifts in range

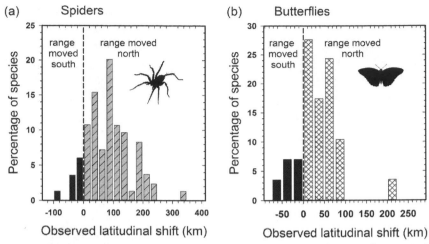

Figure 1.6 Observed latitudinal shifts in the range boundaries of (a) spiders (85 species) and (b) butterflies (29 species) studied over 25 years in Britain. The dashed line marks the point of no range change, and the black bars indicate species that have shifted south, contrary to predictions. (Modified from Chen et al. 2011.)

limits. The simple model for climatic limitation is that geographic ranges for all species should be shifting poleward. But, for example, in an analysis of 764 individual species from a variety of taxonomic groups, Chen et al. (2011) found that 22% of the species moved their ranges in an opposite direction from that predicted by this simple climate change model. One important concept in work on changing climate is to map the rate at which climates are changing in relation to the movement of geographic range. VanDerWal et al. (2013) did this for 464 species of Australian birds over the time period 1950 to 2010. They measured the climatic zone in which each bird species lived. They then mapped the observed shift in this climate zone and compared it to the observed change in the same species distribution from bird observation records over the 60 years. The result was that species were shifting their ranges faster than climate was changing, so they could readily keep up with climate change in Australia. This does not of course mean that if climate shifts become faster this generalization will be correct. While many of the Australian birds were moving in the "correct" direction with respect to climate change, some were not, and these species need additional study.

On a local scale many biological interactions such as competition can affect the distribution of a species. Many plants and microorganisms use chemical warfare to suppress possible neighbors that might harm them.

A well-known example of chemical warfare is the action of penicillin, the secretion of a fungus, on other microorganisms. The soil fungus *Penicillium* excretes this antibiotic to protect itself against bacteria. Humans have simply learned to use this chemical for our own protection against disease. The study of human disease is essentially a study of colonization (by microorganisms) of new environments (people), and thus differs only in scale from the starling's colonization of North America. At some time in our lives most of us owe a debt to the chemical warfare of an antibiotic against some disease organism, and the restriction and elimination of the invading microbe in our bodies. Many plants secrete toxic chemicals that inhibit other plants or the animals that try to feed on them. Most of the spices we use in cooking were evolved by plants to stop herbivores from eating them.

Stream fishes provide an interesting case study in changing geographical distributions because they are constrained by stream geography. In France a survey of the range shifts for 32 stream-dwelling species over a 30-year time frame from 1980 to 2009 has illustrated both the altitudinal changes as well as the upstream-downstream changes. In general, with water temperatures rising, the prediction is that stream fish will tend to move upstream to stay within their temperature zone. Comte and Grenouillet (2013) found that fishes in these French streams shifted upstream on average 14 m in elevation per decade, which averaged 0.6 kilometers in distance per decade. For these streams they found that the rate of range shifting was not keeping up with the temperature changes within the streams, and thus range shifts were lagging behind what is needed to adapt to ongoing water temperature increases.

Mangroves are intertidal trees and shrubs that grow around the Earth along coastlines in tropical and warm temperate areas. Mangroves grow in salt water and are sensitive to cold, so they are a good index of changes associated with ocean warming. Mangrove species have expanded their geographical range toward the poles on five continents over the past half century, at the expense of salt marsh (Saintilan et al. 2014). One common species of mangrove, *Avicennia germinans*, has extended its range along the USA Atlantic coast and expanded into salt marsh as a consequence of lower frost frequency in the southern USA. This genus has also expanded into salt marsh at its southern limit in Peru, and on the Pacific coast of Mexico. Mangroves of several species have expanded in extent and replaced salt marsh where protected within mangrove reserves in Guangdong Province, China. In southeastern Australia, a strong expansion of mangroves into salt marshes is now occurring. These changes are consistent with the poleward extension of temperature thresholds coincident with sea level rise, although the specific mechanism of range extension might be complicated by limitations on dispersal. The shift from salt marsh vegetation to mangrove dominance on subtropical and temperate shorelines will have effects on other species in the intertidal community. Larvae from many species of fish rear in mangrove areas, and, from a practical point of view, mangroves protect shorelines from catastrophic wave action during tsunamis (Alongi 2008).

CONCLUSIONS

The ecological processes limiting the geographic distribution of most species of animals and plants are poorly understood. On the global scale geography and climate are the two main limiting factors. While we recognize well the fauna and flora of different continents, we do less well at the local level to understand, for example, why a particular plant occurs in one woodland but not in an adjacent one. Changes in historic geographic ranges are now being caused by two main processes—human introductions and climate change. The general prediction that in a warming world most species will move their geographic ranges toward the poles is now validated in many cases, but some species do just the opposite and move the "wrong" way for reasons that are not understood. We know from bitter experience that moving species willy-nilly from one continent to another is a serious error without very careful study, and yet we allow plant stores and pet shops to sell species well known to be major pests, should they escape confinement. We know enough to do much better.

CHAPTER 2

POPULATIONS CANNOT INCREASE WITHOUT LIMIT

KEY POINTS

- Populations increase and decline but do not go on increasing for long because of a host of limiting factors. Rarely do populations remain constant, and some decline to extinction.
- The ecologist's job is to stop pest species from increasing and prevent threatened species from declining. There is no universal recipe for achieving these goals and detailed studies are necessary.
- The human population is no exception to this ecological rule, and population growth will stop either by careful forward planning or by undesirable catastrophes.

Humans have always expected the biological world to be in a state of balance so that populations of plants and animals did not change much from year to year. When they did observe great changes in abundance, like a locust plague or an outbreak of malaria, they tended to explain these events by superstitions or curses sent from the gods. Many writers even before Charles Darwin recognized the dynamic nature of populations and some sought ecological explanations for why locusts or mice might be very common one year and nearly absent the next. Over the last century ecologists recognized that once you look into the details you find that populations of plants and animals can increase or decline dramatically for a variety of reasons. The simple idea that Nature is always in a state of balance was recognized to be wrong; the search for a more comprehensive explanation of natural population changes is the subject of this chapter.

The world is finite, and so it is not surprising that nothing can increase without limits. If populations are limited in numbers in the long term, what exactly prevents population growth? The flip side of this question is important to conservation ecologists—what causes populations to decline in abundance and even go extinct? To answer these questions, ecologists break down population changes into four components—births, deaths, immigration, and emigration.

Two processes add animals or plants to a population: *reproduction*, the addition through births or seed production, and *immigration*, the addition through movements into a population. Conversely two processes remove organisms from a population: *deaths* and *emigration*. If we want to find out why a population has increased or declined, we can now reduce this question to a determination of which of the four processes has changed to allow increase or collapse. All populations are subject to natural controls that in the long run balance out births, deaths, immigration and emigration so that population size stays roughly the same. But in the short run, paradoxically, we see precious little evidence of any balance. Birds that are common one winter may be rare the next. Garden pests that are bothersome one summer are nowhere to be seen the following year. The "balance of nature" seems to disappear when we look closely for it. The original balance of nature idea had to be replaced once we discovered that fluctuations in natural populations were commonplace. In the process ecologists have uncovered a much more interesting model of population changes.

Population analysis starts by sorting out births, deaths, immigration, and emigration, and then finding out which components of the environment control these processes. Four broad categories of factors are typically analyzed—weather; resources (food for animals or nutrients for plants); other organisms (predators, competitors, or parasites); and a satisfactory habitat or place in which to live.

SERENGETI UNGULATES

The Serengeti region of East Africa contains a wealth of large mammal species that make the area a global magnet for tourism. Elephants, lions, wildebeest, rhinoceros, giraffes, zebra, antelope, and hyenas—the list goes on and we have not yet even mentioned the 500 bird species, including 34 raptors and 6 vultures. The Serengeti is a World Heritage Site of global significance. One of the most spectacular sights in this region are the migratory wildebeest, herds comprising 1.5 million animals that move seasonally from north to south to take advantage of the rainfall that produces the grass on which they thrive. Figure 2.1 shows the history of the wildebeest (*Connochaetes taurinus*) population in the Serengeti from 1957 to 2009, along with data for the plains zebra (*Equus quagga*) and Thomson's gazelle (*Eudorcas thomsonii*).

By 1977 wildebeest in the Serengeti reached maximum numbers and have fluctuated around 1.3–1.4 million since then. A severe drought in the early 1990s caused high mortality from starvation, and the population recovered by 2000. The ecologist asks what stopped the wildebeest popu-

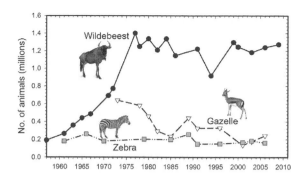

Figure 2.1 Population size of migratory wildebeest (circles), zebra (squares), and gazelles (triangles) in the Serengeti region of East Africa, 1957–2009. (Data courtesy of A. R. E. Sinclair.)

lation increase shown in Figure 2.1, and the answer involves two factors: food supplies and predation. Wildebeest herds can either be migratory or resident. The migratory herds are much larger than the resident herds and they achieve the high numbers shown in Figure 2.1 by avoiding predation losses. By moving away from resident lion populations (which are territorial), migratory wildebeest effectively walk away from their predators and become limited principally by their food supply, which tends to stabilize the population. Occasional droughts produce food shortage by reducing grass growth, and during these droughts wildebeest starve in large numbers only to build up again when the rains return. Wildebeest herds that are resident are controlled more by the predators like lions, and exist at much lower densities than the migratory herds, so they rarely exhaust their food supply (Fryxell et al. 1988).

FOREST INSECT OUTBREAKS

Not all animal populations are stable in numbers like the wildebeest, and the next example illustrates a forest insect population that fluctuates in cycles. The western tent caterpillar (*Malacosoma californicum pluviale*) lays eggs in masses and has gregarious larvae that live in silk tents on tree branches. They have a single generation each year and overwinter as eggs. The larvae fall to the ground when fully grown in midsummer and pupate in the soil to emerge as adults within a few weeks. The adults do not feed and their ecological job is solely to lay eggs for the next generation.

Western tent caterpillars have peaks in population density that occur every 8 to 9 years (Figure 2.2). The pattern of the population cycles is that the duration of the increase and peak phase tends to be 4 to 6 years, and that of the decline phase 3 to 4 years.

Two types of mortality agents produce heavy losses in western tent

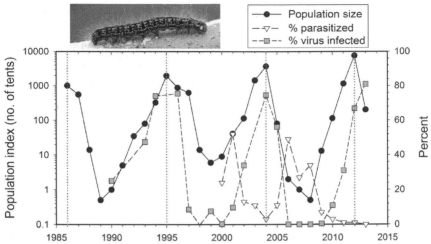

Figure 2.2 Western tent caterpillar on Galiano Island, British Columbia, 1986 to 2013. The population index (black circles) shows an 8–9 year cycle. The percentage of caterpillars parasitized (triangles) rises and falls and is maximal when caterpillar numbers are relatively low. By contrast the percentage of larvae infected by nucleopolyhedrovirus (squares) peaks when the tent caterpillars peak and collapses to near zero when caterpillars are rare. (Data from Myers and Cory 2013.)

caterpillars—insect parasitoids and a virus disease. Parasitoids are insect predators that differ slightly from vertebrate predators in that they lay eggs within insect larvae or pupae, the egg hatches within the host, the parasitoid larva feeds on the host during its larval stage, and then it emerges as a free-living parasitoid when the host dies. Several insect parasitoids kill the larvae of the western tent caterpillar. Some vertebrate predators like shrews feed on the pupal stage. The parasitoids tend to produce variable mortality, often concentrated in the decline and low phase of the tent caterpillar cycles. Viral infections, the second and more significant source of mortality, initiates the population declines of the western tent caterpillar (Figure 2.2).

In addition to the changes in mortality of western tent caterpillars, fecundity, moth size, and fitness also vary with population density. Each female moth lays all of her eggs in a single egg mass that remains on the branch and can be collected and counted after the eggs hatch. Fecundity peaks just before or at peak density and then declines for several consecutive generations so that fecundity cycles in synchrony with caterpillar density. Two hypotheses seem likely to explain this decline in fecundity: (*a*) food limitation and (*b*) sublethal infection of the virus in the caterpillar's

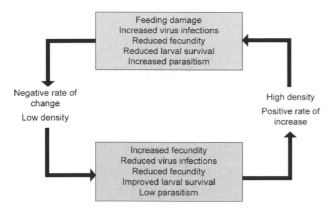

Figure 2.3 Factors stopping increases of the western tent caterpillar and producing cycles in numbers. (From Myers and Cory 2013.)

late larval stages. Infections that do not kill the host caterpillar make the caterpillar grow more slowly and feed less often so it has reduced vitality. On the available evidence, it is not clear whether food shortage or sublethal infections are the main cause of reduced egg counts. Moth fecundity explains approximately 30% of the variation in the rate of population change between years, so both mortality and fecundity are involved in preventing unlimited population growth in these insect pests (Figure 2.3).

POPULATION CYCLES IN GAME BIRDS

Population cycles also occur in some game birds, and in the case of the red grouse (*Lagopus lagopus scotica*) of Scotland the cycle length is 7–8 years. The red grouse is an important game bird in Britain, bringing considerable revenue to Scottish estates, and data from many of those estates illustrate the large cyclic fluctuations in red grouse numbers (Figure 2.4). Research was begun in 1956 to find the reasons for these population fluctuations and to recommend ways of maintaining and increasing grouse numbers. Red grouse live in open moorlands in Scotland, northern England, and Ireland. Their diet is nearly all one food plant—heather (*Calluna vulgaris*). They are not migratory but live their entire life within a small part of the moorland. The red grouse year can be divided into two parts: a period of reproductive gain from April to August, and a period of overwinter loss of birds from autumn to the start of the next spring breeding season.

Overwinter losses in red grouse occur usually in two stages. Numbers decrease sharply in the autumn, remain stable over the winter, and drop sharply again in the spring. Territorial behavior accounts for these two sharp changes. In autumn, family groups break up because of aggression

Red grouse - Scotland

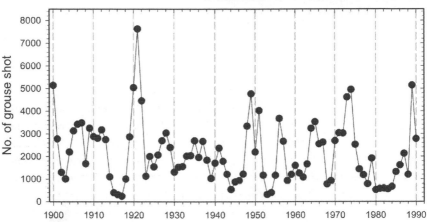

Figure 2.4 Changes in the abundance of red grouse on moor 175 in eastern Scotland, 1900–1990 as indexed by birds shot during the hunting season. This population was strongly cyclic over this time period, with an average cycle length of 8 years. (Data from D. Haydon from Haydon et al. 2002. Photo courtesy of Derek McGinn.)

between the young birds, and two social classes of birds are formed. Territory owners defend a particular area of moor against all comers, and young cocks try to displace territory holders from the previous year. Birds that cannot get territories cannot breed the following spring and are forced to spend most of their time in marginal habitats where heather is scarce. These birds, which form flocks, are then "surplus" birds, excluded by the social system from breeding. Most of the red grouse killed by predators are from these flocks of surplus birds. If a territory owner is killed during the fall or winter, he is quickly replaced by one of these surplus birds.

The second sharp change in the red grouse population occurs in the spring and is controlled by territorial behavior as the breeding season begins. When grouse are very aggressive and take large territories, numbers fall or remain low. When grouse are less aggressive and take smaller territories, their numbers rise or remain high. What determines how aggressive a male red grouse will be, and how big a territory he will take? Two factors seem to be involved. The food available to the mother has a direct effect on the aggressiveness of her offspring. Hens on a high-quality diet produce young that are less aggressive and take up smaller territories. There is also a genetic component to aggressiveness, so that aggressive cocks produce aggressive chicks.

Breeding success is also strongly affected by nutrition. Moors that are artificially fertilized with nitrogen and phosphorus support heather that is richer in protein and mineral content, and female grouse feeding on this high-quality food produce twice as many young as females at poor sites. Good nutrition thus acts in two ways to increase the red grouse population—it increases the production of young birds, and it reduces aggression so that territory size becomes smaller.

The food of red grouse is almost entirely heather, and birds are highly selective in their feeding. They prefer the new shoots of three-year-old heather plants when they have a choice. The nutrient content of the heather is more important than the amount of heather available. A moor covered with a vast quantity of green heather may be inadequate for red grouse if the plants are too old or the soil too poor. One of the recommendations to Scottish landowners that has come from the red grouse research is to improve heather by controlled burning. Every 12–15 years a plot of heather should be burned because by this time the heather has grown too tall for grouse to feed properly. Burnt heather regenerates quickly, and the short nutritious shoots of young heather improve the level of nutrition. In a few years after burning the population size of red grouse goes up. Best results are achieved by rotational burning of long strips 30 meters wide, with older heather left on each side to provide nesting cover. Large fires should be avoided because each grouse's territory must contain nesting as well as feeding areas.

Why do red grouse populations fluctuate? Shooting of red grouse in the autumn has little impact on grouse numbers, and the common sense idea that too much hunting is responsible for grouse numbers fluctuating is not correct. There is a large surplus population available every autumn. Many birds shot in the fall are animals that would die anyway during the winter, and the breeding population of territorial birds is not reduced by hunting in the autumn. If a territory holder is shot by a hunter, one of the grouse from the surplus flock takes his territory.

The maintenance of healthy grouse populations (Figure 2.4) has been due very largely to good habitat management and preventing overgrazing of the moorland by sheep. Sheep browse on heather and thus compete with red grouse for food. On moors where proper burning and heather management are practiced, grouse numbers fluctuate but reach high levels. An interplay between nutrition and territorial behavior sets the limits to red grouse population changes.

Territorial behavior in red grouse operates on a kinship rule—tolerate your neighbors if they are closely related to you and be aggressive toward them if they are not your relatives. This idea can be tested in natural populations by experimentally altering kin structure by moving birds away from their relatives. The predictions from this hypothesis are that close kin structure will permit population increase, and then as crowding for space begins to limit numbers, aggression will rise rapidly and kinship will fall because birds move to other areas or die. Piertney et al. (2008) determined kin structure by mapping all territories on a 140-hectare moor and then using DNA sequences to identify related individuals. The results of their study supported the kin hypothesis as an important cause of population fluctuations in red grouse.

Red grouse require properly managed moors to exist at high densities, and the failure of management during World War II led to the reduction in overall numbers. But within areas of suitable moorland, grouse continued to fluctuate every 7–9 years because of changing social interactions. Predators and hunters kill red grouse, but these factors are not the main drivers of the observed fluctuations.

A GENERAL MODEL OF POPULATION GROWTH

The simplest mathematical model of population growth is the exponential model in which population numbers increase at a constant rate, like a savings account that has a constant interest rate. But endless population increase is not biologically possible, as we have seen with wildebeest and tent caterpillars, so the exponential model has typically been modified by putting a cap on growth and producing an S-shaped growth curve that levels off at some abundance. The idea behind this simple model is that limiting factors grow more severe as population density increases. Ecologists search for these limiting factors that stop growth.

Many ecological factors can affect population increase and collapse—food shortage, predation, disease, parasites, and weather are the major factors. Populations stop growing for a mix of these reasons, and ecologists do not expect to find a general theory that explains for every case the reasons why a particular population stops increasing or declines. We need to investigate each case individually and then draw generalizations for specific groups. For some groups there is an emerging consensus of what factors are likely to be involved.

For large herbivores and carnivores like elephants, whales, wildebeest, lions, grizzly bears, and tigers, population growth is typically limited by their food supplies. For these species there are no predators large enough to check their rate of population increase, with the important exception of humans. While disease or poor weather can ravage populations of many species, these appear most often to be temporary limitations that may be quickly overcome by increased reproduction, as we saw with the Serengeti wildebeest. If any of these large species have effective predators that can eat their young, it is possible that predation can be an important limiting factor. All predators have a limited range of prey sizes on which they can feed efficiently; this means, for example, that wolves cannot survive entirely on a diet of mice while foxes can.

By contrast, small herbivores like rodents and birds have a greater array of limiting factors. Many of these species are territorial, and as we saw with the red grouse, territoriality may limit population growth on a local scale. But either food supplies or predators may also become limiting on a larger scale. In highly variable environments like deserts small organisms are often subject to limitations from rainfall or temperature, acting via food supplies. Small herbivores are often limited by a combination of predation, food shortage, and social interference. Rabbits introduced to Australia entered an ecosystem with few effective predators, and they overran the countryside, being limited in numbers only by food shortage during droughts and by introduced diseases to which they had never been exposed (Saunders et al. 2010). The key to all of these generalizations is that the "devil is in the details," and quantitative data coupled with experimental manipulations are required to see where in the list of ecological processes the important limitations occur. We know too little about the effects of disease and parasites on population changes, and while much research has concentrated on food supplies for animals, nutrients and grazing impacts for plants, and predation on animals, all environmental factors deserve concentrated study.

PLANT POPULATION DYNAMICS

Compared to animals, long-term demographic studies based on plants are scarce. Few data are available on long-lived species like trees, and at the other extreme annual plants have their population size determined

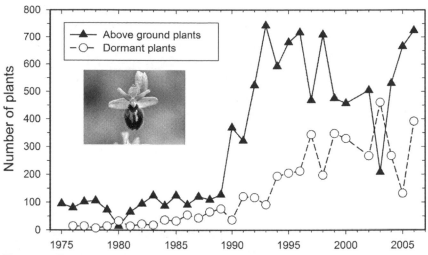

Figure 2.5 The number of emergent and dormant early spider orchids in a 20-by-20-meter permanent plot in southeastern England, 1975–2006. Cattle grazed the study site from 1975–79, followed by sheep grazing. (Data from Hutchings 2010, fig. 1a.)

directly by germination and seed production. One of the best long-term studies was made by Hutchings (2010) on the early spider orchid (*Ophrys sphegodes*), a rare orchid growing on chalk grassland in the Castle Hill National Nature Reserve in southeast England. For 32 years plants in a permanent quadrat were counted each autumn, and all the individual plants were mapped. The early spider orchid is a rare plant, and the nature reserve was managed to protect a large population of this species. The study was carried out from 1975 to 2006, covering two periods of land management, first by cattle grazing (1975–1979) and then by sheep grazing (1980–2006).

The early spider orchid is a short-lived tuberous European orchid of chalk and limestone grassland. It produces a rosette of leaves in September–October and flowers in spring of the following year, from late April to the end of May. Every year, some plants are dormant, and some emergent plants are vegetative. In most years only about 10% of plants produce seeds. The geographic range of this orchid in Britain has decreased by 80% in recent years, so it is now a protected species.

Figure 2.5 shows the changes in numbers of the early spider orchid from 1975 to 2006. The period of cattle grazing was detrimental to this orchid, which began to recover in numbers only after sheep grazing was introduced. Cattle inflict greater mechanical damage on the habitat and vegetation than sheep, causing high mortality and low recruitment in the orchid population. Some grazing is essential in these grasslands to protect the orchid. Grazing prevents more aggressive and taller plant species from outcompeting *O. sphegodes*.

The rapid increase in the orchids after 1989 was explained as a delayed response to the introduction of sheep grazing in 1980. As plant numbers increased, there was an increase in the number of plants remaining dormant, and about 30% of plants remained dormant for periods of 2–4 years. At high density after 1993 plant mortality increased, causing the population to level off at about 750 plants on the 400 m² plot.

Most of these orchids lived for a very short time, only one year. A few plants lived more than 20 years, but most died within one year of appearance above ground. Mortality rates increased in years with high temperatures, and the peak flowering time has moved from late May to early May, a change of about 15 days over the 32 years of study, associated with warming temperatures. Continued climate change could affect the conservation of this rare orchid, and its presence depends on continued disturbance associated with some grazing. Population growth in this orchid was limited by competition for space at high density, so that many plants became dormant and did not flower, and by climate, principally temperature.

HUMAN POPULATION GROWTH

If in the natural world populations of plants, animals, and microbes rise and fall and never increase without limit, perhaps the most glaring exception today seems to be the human population. Figure 2.6 shows the growth of the human population since the year 1500, and three projections to 2100. In 2015 there were 7.3 billion people on Earth. If we accept the fact that no population grows without limits, we can ask which of the four processes involving births, deaths, and movements can operate to achieve limitation. Two of the natural processes for populations—immigration and emigration—can operate for humans only at the local level and not at the global level. We are left with reproduction and mortality.

Throughout most of the last 500,000 years humans have been a "rare" species on the Earth. Human populations rose and fell back again under the combined assault of starvation, disease, and climatic catastrophes added to the self-inflicted losses caused by local and regional wars. Population growth when it occurred averaged less than 0.01% per year. Egypt, for example, had virtually the same number of people in 2500 B.C. as it had in 1000 A.D. Gradual improvements in agriculture, public health, and

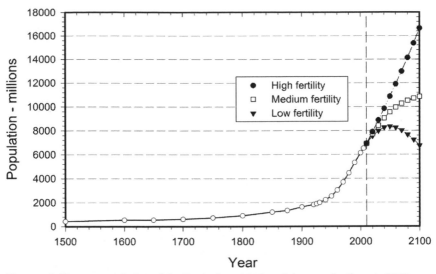

Figure 2.6 Human population of the Earth since 1500 and three projections to 2100 depending on the fertility rate. The high fertility rate is 2.35 children per female, the medium fertility is 1.85, and the low fertility is 1.35. (Source: United Nations 2013.)

housing and a reduction in warfare over the last thousand years have put the human population on a growth spiral that has become extremely rapid during the last 100 years.

The human population cannot increase without limit. Most people accept this fundamental truth, and the problem comes down to how and when the human population will stop growing. We differ from other organisms in having the possibility of imposing our own controls, rather than relying on natural mechanisms of famine, disease, or aggression to set a balance.

To achieve a stable population on Earth we must reach equality between the birth and the death rate. The simplest measure of births and deaths is a tally of the annual number of births per 1000 population and the annual number of deaths per 1000 population. There are two different ways to achieve a stable population: either with low and equal birth and death rates, or with high and equal birth and death rates.

There have been great medical advances during the last 200 years that have reduced the human death rate to a low level, and while the death rate can be reduced even more in many poor countries, for much of the world the death rate is already very low. So the question becomes how to achieve a low birth rate. In general 2.1 children per female is the replacement level

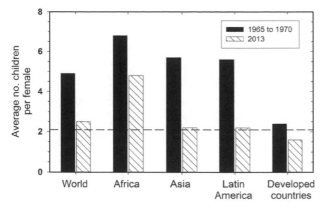

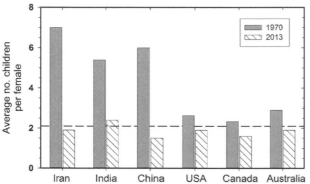

Figure 2.7 Changes in human births per female from 1965–1970 to 2013 in regions of the world and in particular countries. (Source: Population Reference Bureau 2013.)

for births in human populations. Figure 2.7 shows that the birth rates are falling around the world and that many countries are now slightly below the replacement level of 2.1 births per female. The trends in births are encouraging and, if birth rates continue falling, the human population of the Earth could level off or fall somewhat in the next 100 years.

The next critical question is how many people the Earth can support. A variety of approaches have been used to estimate the carrying capacity of the Earth for humans. One promising approach, the ecological footprint, is to recognize that we have multiple constraints because we need food, fuel, wood, water, and other amenities like clothing and transportation. If we express everything in terms of the amount of *land* needed to support each activity and then sum these requirements, we can ask how much land is available in a particular area and how much is available for human use. But this approach also has limits because it is difficult to express energy requirements directly as land areas, and water requirements may be more of a constraint than land areas. It is an index of the carrying capacity of the Earth, and it identifies countries that have a surplus capacity and those in deficit. In 2014 the world as a whole was judged to be in a slight deficit. Australia, Canada, New Zealand, Sweden, and Norway, for example, have more ecological capacity than they are currently using. By contrast, the USA, UK, China, India, and many European countries are using more resources than they have available. If you think of an ecological footprint as a bank account, you can picture the problems of spending more money than you have coming in. It is not a long-term solution for sustainable living.

Other indices of the human impact on the Earth are being developed to quantify specific elements like water use, biodiversity loss, and land degradation. All of them point in the same general direction and suggest that the current size and impact of the human population on the Earth are not sustainable. People can run their bank account down and go into debt, and we can do the same to the natural resources of the Earth, but that is not wise. We should view the Earth as a long-term investment, not as an asset to strip.

CONCLUSIONS

The generalization that populations cannot increase without limits could be called a law of ecology and is a simple recognition that the Earth is finite. The job of the ecologist is to find out which of the processes of births, deaths, and movements set the limits to growth for plants, animals, and microbes, and within that framework determine which ecological factors dictate changes. The majority of research has focused on predation and parasitism, food supplies and nutrients for plants, and climate as key factors stopping population growth. But social and genetic factors are now being recognized as significant limitations for some species and much work on disease is now under way. Landscape factors set other limits on population growth that we will explore in more detail in the next chapter.

CHAPTER 3

FAVORABLE AND UNFAVORABLE HABITATS EXIST FOR EVERY SPECIES

KEY POINTS

- For rare and endangered species we need to turn poor habitats into favorable ones by finding and changing the limiting factors. For pest species we wish to do the opposite, to turn favorable habitats into poor ones.
- Biological control is one way of doing this for invasive pest species and it can work very well but can also fail completely if we cannot find the Achilles' heel of the pest.
- Many examples from natural history tell us what habitats are favorable for a particular species, but we rarely know exactly why this is so. This shortage of knowledge can be important if a species becomes threatened.

Knowing a good naturalist is an enlightening experience. He or she knows where to look for a yellow warbler or a rare orchid, or the best place to fish for rainbow trout. Naturalists have known for centuries that the tapestry of nature is a mosaic of favorable and unfavorable habitats for any particular species of animal or plant, and an ecologist now asks how one can measure the attributes that make one area good and another poor. Ecologists try to find out why one habitat is better than another for two practical reasons. Conservation programs are often devoted to saving endangered species and protecting established species. We can protect desirable species like the woodland caribou or beluga whales only if we know what a favorable habitat includes. Pest control programs work in the opposite manner—now ecologists strive to make habitats unfavorable for the pest species, and only by knowing the species' ecological requirements can we achieve this goal. In each case we search for the Achilles' heel of the species, the weakest part of the species' requirements that can be controlled or manipulated.

Each plant and animal species is affected by all the environmental factors—weather, nutrients, water, other species, and shelter. Continued presence in a region depends upon all these limiting factors being favorable for at least some part of the year, as we saw in chapter 1. For a species to thrive, all the links in the ecological chain of requirements must be present, and the ecologist tries first of all to locate the weak links, realizing that the complete description of the whole chain of environmental requirements may be impossible for the present time. Any environmental factor may provide the weak link.

BIOLOGICAL CONTROL OF PESTS

Biological control has been used for many years to try to reduce the abundance and the impact of introduced pests. In biological control pests are reduced by the introduction of predators, parasites, or diseases, by genetic manipulations of crops or pests, by sterilizing pests, or by disrupting mating through the use of pheromones as sex attractants. Many examples of successful biological control have been described, but there have also been many failures. In successful cases, the habitat of a pest species was changed from good to poor by finding its weak spot.

Diffuse knapweed (*Centaurea diffusa*) is a serious weed of Eurasian origin that was introduced to North America in the early 1900s. It has since spread to over a million hectares of rangeland in western Canada and the United States. Diffuse knapweed is a short-lived perennial plant with seed germination occurring in the spring and autumn associated with rain. Rosette plants develop over the spring and if they reach a sufficient size, they flower in May and June, or remain nonreproductive as rosettes until the next year. Knapweed is a serious rangeland weed because it is poor forage for cows and displaces grasses. Since 1970, 12 species of insects have been introduced for the biological control of knapweed, and 10 have become established; of those 10, 4 are now widespread and abundant (Myers et al. 2009).

The successful biological control of diffuse knapweed was slow; it took 30 years (Figure 3.1). The first three biological control species that were introduced reduced seed production dramatically, but did not kill plants or reduce plant density. Reducing seed numbers for many weeds is ineffective as a means of control because so many seeds are produced there is always a good seed bank even when 90% of the seeds are destroyed. The biological control species that became effective was the weevil *Larinus minutus*, which was initially introduced between 1996 and 1999 and spread to many sites. This beetle feeds on the leaves, stems, and buds of knapweed, and their feeding damage can kill plants, particularly in dry summers.

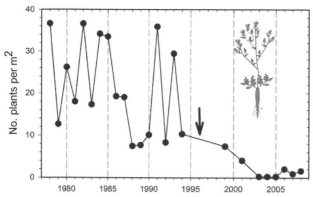

Figure 3.1 The mean density of flowering stems of diffuse knapweed at White Lake, BC. from 1978 to 2008. Three biological control agents introduced in the 1970s and 1980s became abundant but did not reduce plant density. The weevil *Larinus minutus* was introduced in the late 1990s (arrow) with dramatic results. (From Myers et al. 2009.)

The decline in diffuse knapweed density in the dry interior of British Columbia after 1999 has been extensive. Knapweed densities have declined at many locations and along roadsides that formerly were highly infested. This biological control success following the establishment of *Larinus minutus* extends beyond British Columbia to sites in Montana and Colorado. It is a good example of how one can change a favorable habitat for a pest species into an unfavorable one by the introduction of biological control agents.

In the reverse direction there are many examples of how management can change a relatively unfavorable habitat into a better one, leading to population growth. A spectacular example of this comes from the lesser snow goose (*Chen caerulescens*) of North America. These geese are migratory, moving in summer to nest at a variety of tundra sites across northern Canada and Alaska, and then moving back south to overwinter in the southern USA and northern Mexico. Populations of the lesser snow goose have increased 5–7% annually from about 1970 to 2000, when the rate of increase slowed to about 2% per year (Figure 3.2). The rise in numbers coincides with the increased use of agricultural land for winter feeding. Until the late 1940s, snow geese overwintered in narrow bands of salt marshes along the northern coast of the Gulf of Mexico. They began in the 1950s to move north to feed in the rice-growing areas of southern Texas, and once they began feeding on crops of rice and soybeans in the extensive agri-

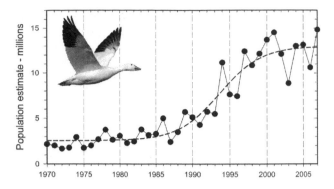

Figure 3.2 Population changes in the lesser snow goose in North America, 1970 to 2006. (From Alisauskas et al. 2011.)

cultural areas away from the coast, the food limitation that existed in the coastal overwintering sites disappeared, the survival rate improved, and populations began to increase.

The increasing abundance of lesser snow geese began to have drastic effects on their nesting areas in northern Canada already by the 1980s. Grubbing for roots in the early spring destroyed large areas of coastal marsh around Hudson Bay, and wildlife managers began to implement plans that would decrease the numbers of geese, such as by liberalizing the hunting season from 1989 to 1997 (1 September to 10 March) as well as allowing hunters 3–4 times the previous daily harvest limit. However, the snow goose populations continued to grow and damage was now extending to the agricultural areas along migration routes as well as on the tundra nesting areas. New regulations specifically intended to reduce population growth of midcontinent lesser snow geese were implemented in 1999. These regulations allowed additional harvesting with several new methods (e.g., no daily harvest or possession limit, and allowing spring hunting along the migration routes). As a result of this additional hunting pressure, each year since 1998 approximately 600,000 geese have been harvested by hunters during the fall and winter seasons across all states and provinces and an additional 320,000 geese in the liberalized spring hunt (Alisauskas et al. 2011). In spite of this increased harvest, it was not large enough to stop population growth. Not enough snow geese could be killed by hunters to stabilize or even reduce the population. Consequently, the population continues to grow and damage both agricultural crops in the USA and tundra nesting habitat in Canada.

Modern farming has removed food limitation for snow geese during migration and on the wintering grounds, so that humans have made a moderately poor habitat that supported a small goose population with low winter survival into a good habitat that supports a much larger population with high winter survival (Abraham et al. 2005). The problem for wildlife managers is that snow geese are now considered a pest by farmers and conservationists who decry habitat destruction.

THREATENED SPECIES MANAGEMENT

For many species that have been declining and are therefore of conservation concern, the path to recovery is difficult until we know the limiting factors. Everyone knows that habitat loss is a major factor in causing species to decline in numbers and become endangered, but the problem is to try to determine how, given the land we currently have available in parks and reserves, to improve survival and reproduction. The next example is from another bird; birds are the best studied of all our threatened species.

The red-cockaded woodpecker (*Picoides borealis*, Figure 3.3) is a species endemic to old growth pine forests of the southeastern United States. It was listed as an endangered species in 1973 due to extensive loss and fragmentation of mature southeastern pine forests. By 1973, the population had declined to <10,000 individuals, which were found in isolated habitat fragments. The red-cockaded woodpecker nests in holes that it excavates in mature longleaf pine trees.

The longleaf pine ecosystem was once the dominant habitat in the southeastern United States along the coastal plain and outer piedmont from Texas to Virginia. When frequently burned (every 1–3 years), the understory communities in longleaf pine ecosystems have among the highest levels of plant species richness of any ecosystem in the world (up to 40 species per m²). Due to widespread timber harvesting, fire suppression, and agricultural development, longleaf pine forests have been severely degraded and fragmented, reducing this forest type to only 3% of its pre–European settlement range. The disruption of the frequent fire regime allows hardwood trees such as oaks to become established. These hardwood trees and their litter alter forest composition, generally degrading the habitat for longleaf pine. Pine-dependent species like the red-cockaded woodpecker disappear as hardwood trees take over these forests (Costanza et al. 2013).

Red-cockaded woodpeckers face two problems. Habitat deterioration is the first and most critical. Without fire, plant succession replaces the pines they require for nesting sites. Logging has also reduced the number of older pine trees that are most suitable for breeding cavities. Second, the conversion of pine forests to agriculture as well as forestry operations

Figure 3.3 Red-cockaded woodpecker, a threatened species now recovering in the southeastern USA. (Photo courtesy of USFWS and Mark Ramirez.)

have fragmented the original pine habitat into a series of islands so that woodpeckers must disperse across inhospitable terrain to reach an unoccupied site.

Two solutions have been applied to this conservation problem. First, fire has been reintroduced into woodlands, a reversal of the historical policy of fire suppression in managed forests. The use of fire in land management has complicated social constraints such as the cost of prescribed burning and a shortage of trained personnel. In addition, there is a potential for damage to human health or property if smoke or fire spread to populated areas. In landscapes that contain a mixture of protected, residential, and agricultural lands, fire use is particularly constrained because of the interface between urban and wild land. Frequent fire by itself seems to be the most useful and least expensive management technique where it can be applied (Steen et al. 2013). The second solution is to provide artificial nesting cavities in pine trees because they are a resource in short supply.

A second problem concerns the social organization of this endangered woodpecker. They live in groups of a breeding pair and up to 4 helpers, nearly all males; this arrangement is called "cooperative breeding." Helpers do not breed but assist in incubation and feeding. Young birds have a choice of dispersing or staying to help in a breeding group. If they stay, they become breeders by inheriting breeding status by the death of older birds. Helpers may wait many years before they acquire breeding status.

From a conservation viewpoint, the problem is that red-cockaded woodpeckers compete for breeding vacancies in existing groups, rather than form new groups. New groups might occupy abandoned territories or move to a new site and excavate the cavities needed for nesting. The key reason they stay in existing groups is the difficulty of excavating new breeding cavities. Because of the time and energy needed to excavate new cavities, typically several years, birds are better off competing within existing territories than building new ones. Dispersal is limited, and nesting cavities are a scarce resource in unoccupied pine forests.

To test this idea, conservation biologists artificially constructed cavities in pine trees at 20 sites in North Carolina. The results were dramatic—18 of 20 sites were colonized by red-cockaded woodpeckers and new breeding groups were formed only on areas where artificial cavities were drilled. This experiment showed clearly that much suitable habitat remains unoccupied because of a shortage of cavities and a lack of dispersal (Walters 1991). Management of this endangered species should be directed not only toward reducing mortality of these birds but also toward providing tree cavities suitable for nesting.

These management actions were applied to the endangered red-cockaded woodpecker population at the Savannah River Site in South Carolina. This woodpecker was rescued from near extinction by a combination of adding artificial cavities for nesting, good fire management, and translocating birds from larger populations nearby. The woodpecker population responded dramatically, increasing from 4 to 99 individuals. Populations are now on an increasing trend across 11 of the southern states (http://www.fws.gov/rcwrecovery/).

The red-cockaded woodpecker is a good example of how a habitat that is currently poor can be made better by manipulating the controlling ecological conditions like fire frequency and limitation of nest sites (Walters 1991).

DISEASE AS A CONTROLLING AGENT

Disease can be effective for reducing a pest species; there are some spectacular examples of how pest populations were dramatically reduced after the introduction of a disease. The most effective diseases are typically ones to which the species in question have not been exposed. We saw an example of this in chapter 2 in the recovery of wildebeest after rinderpest came into Africa (page 15). Another example is the European rabbit, which was introduced into Australia and became a major pest.

The European rabbit (*Oryctolagus cuniculus*) was introduced into Australia in 1859, and within 20 years its population had reached very high densities. Beginning in 1950, the Australian government attempted to reduce rabbit numbers by releasing the myxoma virus. The European rabbit had no prior evolutionary exposure to this virus, whose original host is the South American jungle rabbit (*Sylvilagus brasiliensis*). Biting arthropods, principally mosquitoes and fleas, transmit the virus passively by carrying the virus on their legs.

The myxoma virus causes myxomatosis, a disease that rarely kills South American jungle rabbits but is highly lethal to European rabbits. After its introduction into Australia, it killed more than 99% of the rabbits it infected. Figure 3.4 shows the precipitous crash in rabbit numbers that followed the release of myxomatosis in one area in the early 1950s. Myxomatosis was also introduced to France in 1952, from where it spread throughout Western Europe. In Great Britain, 99% of the entire rabbit population was killed in the first epidemics from 1953 to 1955. This type of extreme mortality is common when pathogens infect new host species, and it can be useful if the host species are pests.

Both the myxoma virus and the European rabbit have been evolving since the virus was introduced into Australia and Europe. More virulent strains of the virus have been replaced by less virulent strains, which kill fewer rabbits and take longer to cause death. Because the host remains alive longer while infected with less virulent strains, the virus has a higher

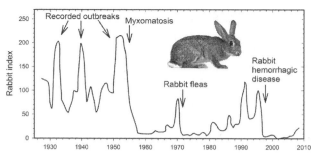

Figure 3.4 Trends in European rabbit abundance in northeastern South Australia. Population fluctuations prior to myxomatosis were reconstructed from rainfall records. Myxomatosis initially reduced rabbit numbers by about 90%. Introductions of European rabbit fleas facilitated the transmission of myxomatosis. Rabbit hemorrhagic disease (RHD) was introduced accidentally in 1997. (After Saunders et al. 2010.)

probability of being spread than more virulent strains. At the same time, rabbits have become more resistant to the more virulent strains of the virus through natural selection.

What impact do these changes in the virus and the rabbits have on the population dynamics of the rabbits? There was a slow recovery of rabbit populations in Australia during the 1960s and 1970s, and an additional insect carrier of the myxoma virus, the European rabbit flea, was introduced in 1970 and rabbit numbers dropped to low numbers in the 1970s. But in the 1980s and 1990s rabbit populations again began to increase, and a search was made for an additional disease to introduce. The virus that causes rabbit hemorrhagic disease (RHD) was first described in China. It spread to Europe and was brought to Australia to study for use as a biological control agent but then escaped to spread in 1997–98 and cause another rapid drop in rabbit numbers. Since 2001 rabbit numbers have been again slowly recovering, and the rabbits have been a local pest on a variety of rare plants (Saunders et al. 2010). Myxomatosis and RHD are good examples of the strong impact that an introduced disease can have on a wild population, turning a favorable environment for the rabbit into an unfavorable one. But the problem is not solved because of the evolution of disease resistance, and as with many pest problem solutions, a stable continuous reduction in pest numbers can only rarely be achieved. Biological control is not always a final solution.

AGRICULTURAL CHEMICAL IMPACTS

Monarch butterflies are an iconic species of butterfly because they migrate each year from Mexico north into the USA and Canada. They are the official butterfly in 7 US states, and consequently are of considerable interest. Since 1994 estimates of monarch populations have been declining in their Mexico overwintering sites (Brower et al. 2012), and conservation biologists are justifiably alarmed (Figure 3.5). This has led to investigations of possible habitat deterioration in both the summer and winter sites occupied by these migratory butterflies.

Monarch butterflies have a multigeneration annual cycle. In the spring, adults that have overwintered migrate north and reproduce in Texas and states to the north and east. Their offspring move farther north into much of the eastern half of the United States and southern Canada, and two to three more generations are produced during the summer (Pleasants and Oberhauser 2013). Adults that emerge after mid-August migrate from the summer breeding range to their wintering grounds in Mexico, where they remain until spring.

Three factors are implicated in the downward trend in the monarch's abundance: (i) the loss of and reduction in quality of critical overwintering habitat in Mexico through extensive illegal logging; (ii) the widespread reduction of breeding habitat in the United States due to the loss of the monarch's principal larval food plant, the common milkweed *Asclepias syriaca*; and (iii) periodic extreme weather conditions.

The most likely explanation for this population decline is that monarch reproduction has been decreasing because of the loss of milkweed, the larval host plant. On the basis of milkweed cardenolide fingerprints, 92% of the monarchs wintering in Mexico had fed as larvae on the common milkweed. Pleasants and Oberhauser (2013) tested the idea that the monarch decline was due to a shortage of milkweed plants. They measured the abundance of milkweeds in agricultural fields and in conservation lands in Iowa from 1999 to 2010. There was an 81% decline in milkweeds on agricultural land over this 12-year period, and a 31% decline in milkweeds on nonagricultural land. Since there is about 4 times as much land in agricultural use as in nonagricultural use in Iowa, the overall loss of habitat for monarch butterflies is about 72%, or about 6% per year.

The critical question is how this habitat loss has affected population changes of the monarch butterflies. By counting the eggs laid on each individual milkweed plant, Pleasants and Oberhauser (2013) were able to calculate how many monarch eggs could have been laid in the summer habi-

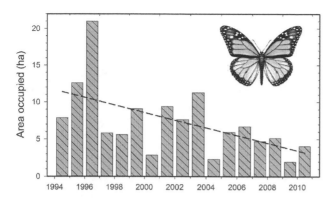

Figure 3.5
Monarch butterfly overwintering area in central Mexico from 1994 to 2011. On average the overwintering area occupied declined 0.5 hectare per year. (Data from Brower et al. 2012.)

tat range from 1999 to 2011. Monarch egg production over the 12 years declined about 81%, and this decline in reproductive potential mirrors the collapse in the overwinter population arriving in Mexico during the next autumn. It would have declined much less if monarchs had laid more eggs on the remaining milkweed plants in nonagricultural area, but they did not do so. The count of eggs per milkweed plant in nonagricultural habitats over this time was essentially constant.

Why did these losses of milkweed plants occur? The suspected connection is with agricultural herbicide use. The monarch decline coincides with the increased use of glyphosate herbicide in conjunction with increased planting of genetically modified glyphosate-tolerant corn (maize) and soybeans. Glyphosate is used to kill weeds in the agricultural fields and kills milkweed plants. Glyphosate was not used very much for weed control until the late 1990s, when glyphosate-tolerant crops were planted. The use of glyphosate in agriculture increased dramatically since 2000.

There have been losses of forest habitat in the overwintering sites in Mexico due to logging and illegal tree harvesting, and this is a second potential explanation for the population reduction of monarchs. But in this case the greatest amount of the decline in monarch numbers is accounted for by the egg losses in the summer range. Both winter and summer habitat must be protected for migratory species, and Mexico is pushing ahead with strong habitat protection for overwintering monarch populations.

The collapse of monarch butterflies illustrates how agricultural modifications can strongly affect wildlife for better or worse. The important point is to appreciate and study these kinds of effects before major changes are made to how we grow crops. This lesson has yet to be learned, unfortunately.

SOIL CHEMISTRY AND AGRICULTURAL PRODUCTION

Agriculture has provided us with numerous examples of areas that were not favorable for crops because of micronutrient deficiencies in the soil. The ironic aspect of this soil limitation has been that large areas have been set aside for national parks and reserves because the land was not deemed suitable for crops. Much research by agricultural scientists has found out why these limitations exist and in the process has achieved an ability to change poor habitats for crops into good ones.

Australia has numerous micronutrient-deficient soils because it is an ancient continent with highly eroded soils (Orians and Milewski 2007). Numerous attempts to grow crops in low-rainfall areas of Australia have failed until experimental work was carried out to determine the causes of crop failures. Many of these failures were due to the shortage of micronutrients in the soil—boron, copper, iron, chloride, manganese, molybdenum, and zinc. These soil nutrients are needed in minute amounts for plants and animals, and are often called trace elements. Many of Australia's plants are adapted to live on very low-nutrient soils, but when agricultural crops were introduced many problems arose. Dryland ecosystems around the world are particularly susceptible to micronutrient deficiencies or in a few cases micronutrient excesses (Ryan et al. 2013). Many of the agricultural experiments done on micronutrients illustrate well the simple manner by which poor soils for crops can be changed into good soils by the addition of micronutrients. Two examples will illustrate these global agricultural problems.

Wheat grown in many dryland soils in Australia requires nitrogen fertilizer to obtain maximum yield, but Lipsett and Simpson (1973) discovered that the addition of molybdenum had a stronger effect on wheat growth and yield and that in this particular instance soil fertilization with nitrogen reduced the yield (Figure 3.6).

The successful development of modern agricultural industries around the world could not have been achieved without the discovery and correction of trace element deficiencies in crops, pastures, horticultural tree and vegetable crops, plantation forests, and livestock. Most of this research was done during the last century, and the low–trace element status of many agricultural soils has been demonstrated and often multiple deficiencies have been identified. Their amelioration with trace element fertilizers led to the development of large tracts of infertile land for agricultural production; spectacular increases in yields and the quality of harvested products; and marked improvements in livestock productivity. These re-

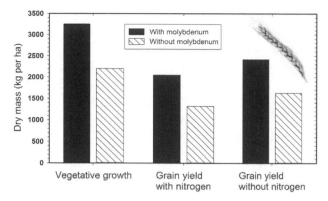

Figure 3.6 Effect of nitrogen and molybdenum fertilization on wheat production in southeastern Australia. Vegetative growth refers to total plant biomass, including stems and grain. Grain yield refers only to the grain harvested. (Data from Lipsett and Simpson 1973.)

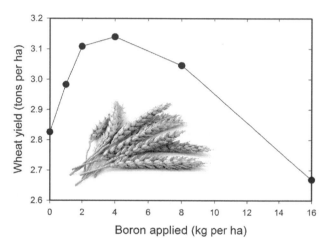

Figure 3.7 Wheat yield in rain-fed soils in the Punjab Province of Pakistan, 2002–3, in relation to addition of the micronutrient boron. A maximum improvement of 11% in yield occurred at 2–4 kilograms boron per hectare. Too much boron produces boron toxicity. (Data from Rashid et al. 2011.)

search results have largely been incorporated into best-practice agriculture around the world, and have resulted in a global experiment of how to change poor habitats into good ones for crop production.

Of the soil micronutrients, at the global level, zinc and boron are the most important in terms of the areal extent of deficiency and are serious constraints to crop yields. In Pakistan, for example, the identification of deficiencies of boron and zinc in wheat, cotton, and rice has led to cost-effective crop responses to these limiting micronutrients, and consequent farmers' adoption of applying boron and zinc for these crops (Rashid et al. 2011). Figure 3.7 illustrates one example from Pakistan in which adding small amounts of boron in fertilizer increased wheat yield 11%. Boron availability in soils is variable spatially, and spatial mapping of soils is needed to specify areas likely to suffer from micronutrient deficiencies.

Many millions of people in the world suffer from an insidious form of hunger known as micronutrient malnutrition. Deficiencies of zinc and iron afflict more than one-third of Earth's human population—women and children in resource-poor families are the most affected (Fageria et al. 2012). More than 60% of the world population suffers from iron deficiency, and over 30% of the global population has zinc deficiency (Rawat et al. 2013). All micronutrients are essential parts of various enzymes needed for carbohydrate metabolism. Iron is also a critical component of haemoglobin. Micronutrient-deficient soils produce micronutrient-deficient feeds and foods, which, in turn, cause animal and human malnutrition. The complex interplay of soils, crops, and human nutrition shows in too graphic detail the important problems that have to be resolved to convert poor agricultural systems to good ones. There is increasing research on these problems for the human health issues involved.

CONCLUSIONS

Good and poor habitats exist for every species on Earth, and for species conservation or pest management it is critical to know exactly what factors dictate how good or poor an environment is. Management has been successful in some cases of changing habitats in a desirable direction by adding or removing predators or diseases, or changing the chemistry of soil or water. Unfortunately humans have been equally successful in making habitats unsuitable by changing the same ecological factors either inadvertently or in carefully planned ways that had unanticipated consequences. If there is a general message from the good and bad stories illustrated in this chapter, it is that we should be more risk-adverse with how we change natural ecosystems until adequate studies have been completed.

CHAPTER 4

OVEREXPLOITED POPULATIONS
WILL COLLAPSE

KEY POINTS

- Any harvested population must decline in abundance, and the losses due to harvesting may be compensated for by increased growth or reproduction or decreased natural mortality. Overharvesting can lead to the extinction of the resource.
- Overharvesting occurs more frequently in common property resources unless there are strict, enforceable regulations. In many fisheries, private or communal ownership and control can prevent overharvesting.
- Humans typically harvest the largest individuals, which can select against large body size and rapid growth rates. These changes in body size can have a genetic component and result in artificial selection for less fit genotypes, to the long-term detriment of the species.

From the earliest days humans have eaten both animals and plants and this harvesting has intensified as the human population expanded. When human populations were scattered at low density, harvesting was light, and no problems arose because groups could move to a new place, as yet untouched. But when new technologies were developed and the human population increased, intensive harvesting became more common and larger areas were affected, with the end result that humans have now changed the face of the Earth. The destruction of forests in the Mediterranean region was completed by Roman times. The Cedars of Lebanon are no more, exploited for 5,000 years for their valuable wood, used by the Egyptians, the Babylonians, the Persians, the Phoenicians, and the Romans for shipbuilding. In spite of these early problems with overharvesting, only during the last hundred years have scientists attempted to analyze the overexploitation problem in a quantitative ecological framework.

By the late 1800s people began to notice examples of overfishing in some marine fisheries. In 1885, the Dalhousie Committee was set up in the United Kingdom to investigate the alleged depletion of fish stocks by the trawl net and the beam trawl. The committee was unable to provide an answer to the problem because no ecological data were available on the fish stocks of the North Sea, and so they recommended that the government begin to collect adequate fish statistics and start scientific work on fish population dynamics.

A SIMPLE MODEL OF FISHING

At the close of the 19th century, fishery scientists began to sort out the pieces of the puzzle that were needed to analyze the overfishing problem. The first breakthrough was the discovery that the age of many fish can be read from growth rings on the scales, first discovered by Hoffbauer in 1898. This discovery meant that scientists could now determine the age composition of a fish population. In 1918 Huntsman in Canada and Baranov in Russia independently discovered that fishing changed the age composition of a fish stock. Fishing removes the older fish, which are also the larger fish, and when a fishery is just starting up the catches of these larger fish may be spectacularly good, providing a signal that overharvesting may be occurring.

In 1931 the British scientist E. S. Russell published a theoretical analysis of "overfishing" that set the exploitation problem in a clear light. For a fishery, interest usually focuses on the weight of the catch rather than on the number of fish, and so Russell analyzed the weight dynamics of an exploited population (Figure 4.1). Two factors decrease the weight of the fish stock during the year—natural mortality and fishing mortality. Similarly, two factors increase the weight of the stock—recruitment of young fish and growth of older fish. Recruitment in exploited populations is a measure of the number of young organisms entering the harvested population. The whole process is thus quite simple:

To balance the fish population, Russell pointed out, gains must equal losses, so that:

Recruitment + Growth = Natural mortality + Fishing mortality

A critical problem for Russell now appeared. Before fishing began, recruitment plus growth, on the average, must have been equal to natural mortality if the population was stable. Now, with exploitation, this equation becomes unbalanced and the total weight of the fish population be-

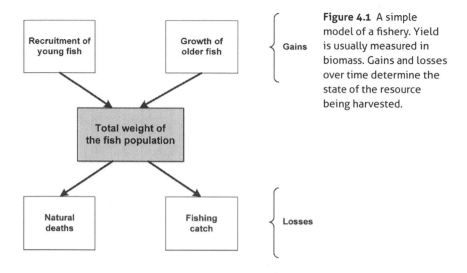

Figure 4.1 A simple model of a fishery. Yield is usually measured in biomass. Gains and losses over time determine the state of the resource being harvested.

gins to fall until either recruitment or growth goes up, or natural mortality goes down. Unless one or more of these three elements changes, the exploited population will decline to extinction.

A LABORATORY TEST OF FISHING THEORY

Laboratory studies of insects and small fishes have been particularly useful in analyzing the basic principles of harvesting theory. One small fish that can be raised easily in small aquaria in the lab are guppies (*Lebistes reticulatus*). Silliman and Gutsell (1958) maintained two populations as unmanipulated controls and two populations as experimental fisheries subjected to a sequence of four rates of fishing (Figure 4.2). Populations were counted once each week and "fishing" was done every third week.

Unfished control guppy populations reached a stationary plateau by week 60 and remained there until the end of the experiment in week 174 (Figure 4.2). Fishing at 25% once every three weeks reduced the experimental populations to about half of the biomass of the controls. Reduction of the fishing to 10% increased both experimental populations to about 70% of the control biomass, and the imposition of 50% fishing in week 121 caused a decline in population size to about 20% that of the controls. A fishing intensity of 75% every third week was too great for these fish to withstand, and the experimental populations were driven to extinction by "overfishing."

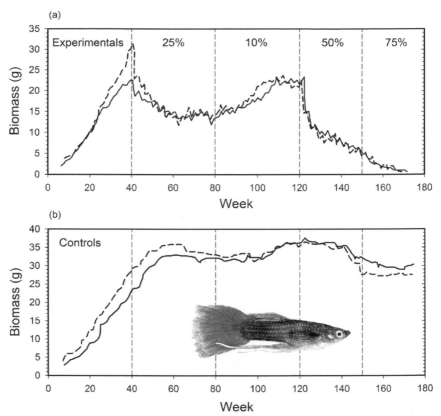

Figure 4.2 Population biomass changes in guppies maintained in the laboratory. (a) Two experimental populations subjected to harvesting after week 40 at the indicated rates. (b) Two control populations that were not exploited. (Data from Silliman and Gutsell 1958.)

These experiments on guppies and many other experiments on wild populations illustrate four principles of exploitation:

1 Exploitation of a population reduces its abundance, and the greater the exploitation, the smaller the population becomes.
2 Below a certain level of exploitation, populations are able to recover and compensate for removals by surviving or growing at increased rates.
3 Exploitation rates may be raised to a point at which they cause extinction of the fishery.
4 Somewhere between no exploitation and excessive exploitation is a level of maximum yield.

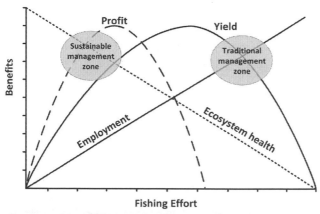

Figure 4.3 Two definitions of maximum yield to a fishery—yield in biomass (solid curve) and yield in dollars (dashed curve) in relation to fishing effort. The potential benefits in employment and ecosystem preservation are shown schematically. Two zones are identified in which the two objectives of management are not compatible. (Modified from Hilborn 2007.)

The problem now becomes this: How does one determine the level of maximum yield for a natural fish population? Maximum yield can be measured in two ways—in biomass of fish, or in dollars (Figure 4.3). The traditional management plan has maximized employment with heavy fishing effort and low profits. A more sustainable management plan would reduce fishing effort and maximize profit and ecosystem health. The practical problem is that it is difficult to move from the traditional goal to the sustainable goal because of social factors concerning employment. The end result is too often overfishing and the collapse of the resource.

THE TRAGEDY OF OVERFISHING:
THE ATLANTIC COD OF EASTERN CANADA

The Atlantic cod (*Gadus morhua*) is a marine fish that occurs in cool northern waters off eastern Canada, living in near-shore areas and out on the continental shelf to a depth of 600 m. Cod have played a major role in the early colonization of North America by Europeans. When John Cabot came from England to Newfoundland in 1497 he found the sea "swarming with fish—which can be taken not only with nets but in baskets let down with a stone." Basque fishermen from northern Spain had preceded Cabot to the Grand Banks off Newfoundland to catch cod, salt them, and carry them back to Europe. Salted cod was a delicacy in Europe during the 15th and 16th centuries, and the need to dry cod on land and salt it drove the

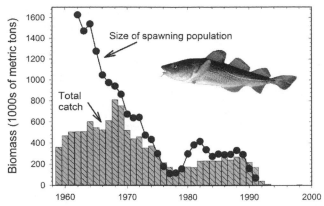

Figure 4.4 History of the fishery for the Atlantic cod (histogram) off Newfoundland from 1959–1998, and the estimated biomass of the spawning population (line). The fishery completely collapsed from 1989 to 1992, when it was closed and it remains closed in 2014. The size of the remaining population after 1992 is too small to show on the graph. (Data from Fisheries and Oceans Canada 1999.)

settlement of the northeastern part of North America. Since Cabot's time the Atlantic cod has been the dominant commercial species of the Northwest Atlantic. Now it borders on extinction, a victim of overfishing, and the collapse of the cod fishery has been a social, economic, and ecological disaster for the people of Newfoundland.

The cod fishery operated sustainably for nearly 500 years. During the 1600s the annual catch of cod was about 100,000 metric tons per year, and this rose as high as 200,000 tons in the 1700s due to the demand in Europe for salted cod. During the 1800s the catch ranged from 150,000 tons to 400,000 tons annually. Until 1900 all the cod was salted and dried. After 1900 fishing boats became larger with more efficient nets, and the efficiency of the fishing fleet continued to increase due to technological enhancements. Frozen fish now replaced the older methods of preservation, and cod fishing intensity continued to grow. During the 1950s about 900,000 tons of cod were harvested in the Northwest Atlantic, and this increased to 2 million tons in the 1960s. Figure 4.4 shows the harvest of Atlantic cod in Canadian waters from 1959 to 1998. In 1977 Canada extended control over coastal fishing out from 12 to 200 nautical miles, and most of the cod fishery off Newfoundland came under Canadian management. Canadian management has been a disaster. Spawning biomass of cod in 1991 was 4% of what it was in 1962. The cod fishery collapsed and the dominant commercial fishery of the Northwest Atlantic was closed in 1992, throwing 35,000 Newfoundlanders out of work. It is still closed in 2014 and little recovery of the cod population is evident. What went wrong with the management of this fishery?

Two major scientific errors occurred in the management of Atlantic cod stocks during the 1980s and early 1990s. First, the estimates of the size of the cod stock were far too high. Cod stocks are estimated by measuring the catch in a series of fishing surveys off the coast. Cod were not randomly spread over the fishing area but were concentrated in high-density aggregations; because of this, the estimates of population size were too large. Second, the mortality rate of cod from fishing was grossly underestimated because there are sources of mortality that are not measured by fishery scientists. Young cod that are too small are illegal to sell, and this incidental catch ("bycatch") is usually discarded at sea, causing mortality that is due to fishing but was never measured; because this bycatch did not contribute to the yield to the fishery, it wasn't counted in the harvest biomass. As the abundance of older legal fish was reduced, more and more undersized illegal fish would be caught and discarded. The size of this discarded catch can be enormous, with reports from some fishing trawlers of having to catch 500,000 cod and discard 300,000 undersize fish to get 200,000 legal-sized cod. This increase in mortality rates of young fish would impact directly on how many cod reached adult age of 6–7 years and begin to reproduce.

The Atlantic cod collapse has devastated the economy of Newfoundland and cost the Canadian taxpayers at least $4 billion. The recovery of the Atlantic cod will take decades, yet there is continuing political pressure to reopen the fishery. There has been limited evidence of stock recovery during the last 20 years, and cod are only slowly increasing in the absence of a commercial fishery. Fishery scientists estimate that it will take at least until 2030 to have sufficient population recovery to support a normal fishery for cod. The next 15 years look bleak for the fishing industry of Newfoundland, and a fishery that was sustained for 500 years has been destroyed in our lifetime.

Because the fishing industry operates in both a social and an ecological framework, it is subject to political conflicts that confront ecological reality. These conflicts are often not resolved and the fishery suffers. Two forces drive up the fishery's harvest to unsustainable levels. In good years the profits are high and more investments in better boats and better nets are made, and the fishery becomes more efficient. In poor years the government is asked for subsidies to maintain employment and a high harvest rate is maintained when it should be reduced. The net result is to drive the system toward a collapse (Ludwig et al. 1993).

The collapse of the Atlantic cod fishery is an illustration of *the tragedy of the commons*. "The tragedy of the commons" was the term coined by Garrett Hardin in 1968 to describe the overexploitation of resources that are open to anyone to use. Whenever a resource like a marine fishery is held in common by all the people, the best policy for every individual is to overharvest the resource or "beggar your neighbor." There can be no reason to stop harvesting at some optimum point because you as an individual can always make more money by overharvesting, and if you do not overharvest, your neighbor will. This overexploitation of common property resources can be averted only by some form of regulations that restrict harvesting, or by converting a common property resource to a private resource through private ownership. Social control of harvesting is required for all large-scale fisheries, and for this reason good resource management is a creative mix of ecology, economics, and sociology.

THE TRAGEDY OF OVERFISHING: WHALING IN THE SOUTHERN OCEAN

The exploitation of whale populations in the Southern Ocean is a classic example of how overharvesting has long-term effects on populations. All commercial whaling has been stopped for more than 25 years, and most whales are now protected. The large whales comprise 10 species divided into 2 unequal groups. The sperm whale was the only toothed whale hunted commercially. The other nine species were all baleen whales, which have bony plates (baleen) in the roof of the mouth. Baleen whales are filter feeders whose principal food in the Antarctic is krill (85 species of shrimp-like crustaceans) and other plankton.

The history of whaling is characterized by a progression from more valuable species to less valuable species as stocks of the original targets were reduced. Modern whaling dates from 1868, when a Norwegian, Svend Foyn, invented the harpoon gun and the explosive harpoon. In about 1905 whalers pushed south into the Antarctic and discovered large populations of blue whales and fin whales. Blue whales dominated the catches through the 1930s, but by 1955 few were being taken (Figure 4.5). Attention was turned to the fin whale, originally the most abundant whale in the southern oceans. Fin whale numbers collapsed in the early 1960s. Sei whales were ignored as long as the bigger species were available and were not

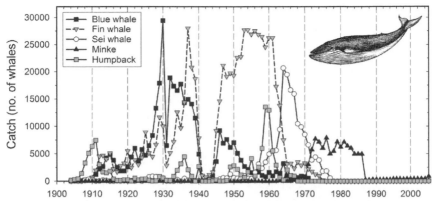

Figure 4.5 Harvest of baleen whales in the Southern Hemisphere, 1910–1977. The usual lengths of whales in the commercial catches were: blue, 21–30 meters; fin, 17–26 meters; sei, 14–16 meters; humpback, 11–15 meters; and minke, 7–10 meters. As each species was overharvested, whalers moved to catching the next largest species. (Data from FAO Fishery Statistics and Allen 1980.)

harvested until 1958. Sei whale catches were restricted after 1972 by the International Whaling Commission to prevent the collapse of these populations. All whale harvesting has historically followed the tragedy of the commons, sustained overharvesting of the resource until an international treaty was adopted to control exploitation.

The present management of whales is directed to measuring the recovery rate of the depleted whale populations. Paradoxically, most of the data we now have on whales have come from whaling operations, and now that commercial whaling has stopped, additional research has to be mounted to monitor how whale populations respond. Whale populations change slowly, and even 10 years is a short time to estimate accurately a population's response to protection from exploitation.

The principal food of the baleen whales, krill, is now being commercially harvested in the Antarctic. Krill are on average about 6 cm long and weigh about 1–2 grams. Krill are so abundant in Antarctic waters that they have been considered for potential harvest for many years. Estimates of the sustainable harvest for krill are extremely large, nearly equal to the total production of all other fisheries on the planet. Commercial harvesting began in the 1970s, and has been hampered by the remote location of the Antarctic and by processing problems. Krill are used for feed supplements in aquaculture and for omega-3 krill oil supplements for the nutrition industry. One of the emerging conservation problems of the southern oceans is to estimate the impact of krill harvesting on the recovery of whale populations in the Antarctic (Grant et al. 2013).

The history of whaling and the decline of the large whales from overharvesting is another unfortunate example of the pursuit of economic gains to the detriment of ecological integrity in the oceans. Fortunately, the large whales are now on the slow road to recovery.

ARE CURRENT MARINE FISHERIES SUSTAINABLE?

There is considerable controversy now about the sustainability of marine fish harvests. Because the abundance of marine fish stocks is poorly known, estimates of the number that are overharvested vary between observers who rely on different types of data (Pauly et al. 2013). The last assessment of marine fish stocks was done in 2006, when complete data on the biomass of the catches were available. A total of 15% of marine fisheries were overexploited, while 13% had collapsed, 48% were fully exploited, and 24% were still developing. The important point is that the marine fishery resources of the Earth are operating at close to maximum capacity, and while some fisheries could be expanded, most were nearly fully utilized even in 2006.

There are two strategies for helping overfished stocks to recover. The first and most difficult is to reduce the fishing pressure directly. This is made difficult because of social objections to the loss of jobs and income, and because of illegal fishing that cannot be easily stopped. A second general strategy is to impose protected areas, or "no-take" zones, on the resource. This is a bet-hedging strategy in which we reduce the catch (an economic cost) for the benefit of a reduced risk of catastrophic collapse of the fishery. This strategy has been discussed particularly for marine fishes, but it will work also for freshwater fishes. The idea of a protected area in the aquatic realm is equivalent to the idea of national parks on land. As marine protected areas, they are most useful for bottom-dwelling fish that inhabit large areas of the ocean floor, and are nonmigratory. The idea is simple: set aside a large enough area of a "no-take" zone to ensure that the stock will remain at >60% of carrying capacity over a given time horizon (e.g. 20 years). Fishermen could harvest at a specified rate outside the "no-take" zone but would not be permitted to fish inside this protected area. The details of how to achieve these simple goals need to be worked out for each fishery, and the detailed trade-off of costs and benefits needs to be measured if the protected area strategy is to obtain practical support among fishermen. By defining and protecting marine parks that are no-fishing zones in the ocean, it is possible to prevent overexploitation of many marine aquatic resources.

Marine reserves are a relatively recent concept in fisheries management, and the available data show that they work very effectively to increase populations of organisms in the protected area as well as in adjacent areas. Fish biomass inside the reserves was up to twice the level of that outside the reserves, and the value of a reserve increases with time. Enforcement is key to stop poaching inside reserves. Larger marine reserves are also better than smaller reserves because increasing a fish population from 10 to 20 in a small reserve has less of an impact than increasing the population from 1000 to 2000 in a large reserve. Larger fish like tuna that move extensively cannot be protected in small reserves; larger reserves are needed (Edgar et al. 2014).

A protected area strategy for marine fishes is clearly an important means of preventing overfishing. To convince managers of the value of a reserve, a long-term monitoring program to measure the magnitude of the population changes that may not be apparent in the short term needs to be implemented. The idea of marine reserves is an important new strategy for trying to prevent the kinds of disasters we have seen in the Atlantic cod fishery and in the history of whaling during the last century.

MIGHT HARVESTING BE GENETICALLY SELECTIVE?

Fisheries typically harvest the largest individuals first, and hunters often search for the mountain sheep with the largest horns. Fishery and wildlife ecologists have become concerned that this type of harvesting could lead to genetic selection in favor of smaller, less fit individuals. Identifying trends in harvested fish and wildlife is relatively easy, but it is more difficult to pin down why size might be changing. There are five primary hypotheses—intensive harvesting, genetically selective harvesting, sociological selection for smaller animals, climate shifts, and habitat changes. Monteith et al. (2013) analyzed the records of trophy hunting specimens submitted to the Records of North American Big Game, established by the Boone and Crockett Club in 1932, to look for trends in the largest animals submitted for evaluation each year. They analyzed data from 22,034 trophy specimens of 9 species of antlered game (e.g., moose, deer) and 8 species of horned mammals (e.g., muskox, mountain sheep) collected from 1900 to 2008. Fifty-six percent of their comparisons showed a significantly negative trend over time (Figure 4.6). They suggested that their data were

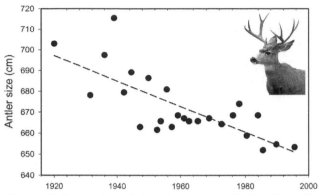

Figure 4.6 Temporal trends in antler size of mule deer (*Odocoileus hemionus*) trophies recorded in Records of North American Big Game during the past century. Antler size is a complex set of measurements made on antlers. From these data, antlers from trophy mule deer are declining in size on average by about 6 cm per 10 years. (Data from Monteith et al. 2013.)

most consistent with the two harvesting hypotheses, but they were unable to determine if harvesting was exerting genetic selection for smaller size in the mammals hunted for trophy antlers or horns.

All intensive harvesting tends to take the largest animals first, so that in general the average size of the fish or wildlife harvested will tend to decline over time. The question then becomes whether this size change (see Figure 4.6) is simply due to the harvesting, so that no genetic selection is being carried out, or whether the genetics of the remaining population is somehow impoverished so that animals that grow slowly are selected for and those that grow rapidly are being selected against. A simple test for this would be to stop harvesting for a few generations to see if the size distribution would revert to larger fish or larger mammals. But this test is very difficult to do with wild populations because the economics of harvesting interferes with experimentation.

Another way to investigate this question is to do selection in the laboratory. Van Wijk et al. (2013) carried out a selection experiment on laboratory populations of the Trinidadian guppy, *Poecilia reticulata*. They selected for large and small males for three generations. Male guppies stop growing when they reach maturity, so adult size can easily be measured. Average male length increased by 7.5% in lines selected for large size, and decreased by 6.5% in lines selected for small size, relative to the unmanipulated stocks (Figure 4.7). Because the genome of the Trinidad guppy is well described, Van Wijk et al. (2013) were able to test for possible genetic selec-

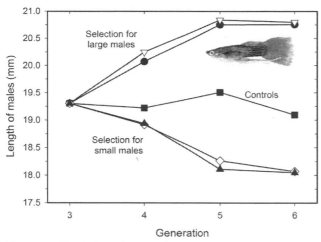

Figure 4.7 Trends in male guppy length for selection lines and the control over three generations of selection for large and small body size in laboratory populations. Two generations were raised in the laboratory to standardize the growing conditions and food supplies, and then selection was applied. (Data from Van Wijk et al. 2013.)

tion produced by the selection for large and small body size. They found strong evidence for selection on individual genetic loci using 6 different methods of genetic analysis.

The bottom line is that body size in many vertebrates is highly heritable; thus, strong selection for larger individuals could potentially change the genetics of the exploited population (Allendorf and Hard 2009). How much these kinds of changes operate in natural fish and wildlife populations is a question being actively pursued.

CONCLUSIONS

All harvested populations of animals and plants are reduced in abundance and if overharvested can be driven to extinction. The key is to find the optimal level of harvesting that maximizes yield over time, and this has been difficult in many harvesting situations because overharvesting can be beneficial to humans in the short term. Social and political factors interact with biological assessments in all harvesting operations, and often short-term thinking or inadequate data produce the economic and social disasters associated with overharvesting of fish, wildlife and plants. Harvesting can be genetically selective, leading to long-term artificial selection of less fit organisms; sustainability in both the short and long term must be an explicit goal of all harvesting.

CHAPTER 5

PLANT AND ANIMAL COMMUNITIES CAN RECOVER FROM DISTURBANCES

KEY POINTS

- All ecological communities are subject to natural disturbances from weather, fire, windstorms, and in some cases volcanos. Human-generated disturbances pose new challenges for recovery.
- Disturbances change the abundance of species in an ecological community, favoring some but not others. These effects are species-specific and community-specific.
- There is little general theory to allow prediction of recovery rates for different disturbances, and careful monitoring of case studies is producing a set of guidelines for the expected rate of recovery for different communities.

Populations of plants and animals do not exist all by themselves but in a mix of other species populations, and we call all the animals and plants in a habitat an ecological community. Communities are the most obvious ecological units to naturalists because we see them every day—a pine forest, a lake, a sagebrush desert, the rocky seashore—and one of the major jobs ecologists have is to try to understand how these communities work. Communities commonly contain thousands of species even if we ignore all the microbes. How do all these species interact to produce the world that we see? We can see how something complex works by taking it apart as an engineer might. This approach can be applied to a biological community by studying the populations of each of the component species. But this is not the only way. Another way to see how something works is to cause some disturbance and watch what happens. In addition to natural disturbances from fires, landslides, windstorms, and volcanos, humans have been disturbing natural communities on an increasing scale in recent years; oil spills, water pollution, logging, and pesticide treatments disturb ecological communities.

HOW DISTURBANCES CAN AFFECT COMMUNITIES

A *disturbance* in ecology is any discrete event that disrupts the plant or animal community structure and changes available resources, substrates, or the physical environment. Disturbances can be destructive events like fires or landslides, or environmental fluctuations like severe frosts. Disturbances like fire operate as a pulse, and the disturbance is over and finished quickly. Other disturbances, like a prolonged drought, operate as a continuous press, slowly pushing in one direction. The notion of what is "normal" for the community is excluded from the ecologists' view of disturbance (in contrast to the everyday use of the word *disturbance*). This is an important change of focus that has implications for conservation and land management. We cannot assume that plant and animal communities in the "good old days" were "normal" and had no disturbances, and that the job of conservationists or land managers is to get back to what the community was like in the "good old days" before human disturbances. Disturbances are part and parcel of community organization and are always occurring. For some communities, disturbances are frequent, but in others disturbances are rare.

Disturbances can affect ecological communities in many different ways depending on their strength and frequency of occurrence. In general, ecologists have considered communities subject to disturbances as recovering slowly back to the original community through a process of succession (Figure 5.1). But if several disturbances hit a community at the same time or in rapid succession, the community may not be able to recover; it will be pushed into an altered state. These disturbances can have particularly severe effects on communities already stressed by human impacts of pollution or climate change.

The simplest model of disturbance is the succession model (Figure 5.1a) in which communities recover over some time period to more or less their original composition. This simple model assumes a uniformity of nature that is now considered rare in a world subject to climate change and human interventions in agriculture and forestry. Consequently, many plant and animal communities are now thought to be more similar to Figure 5.1(b) and (c), and the problem has shifted to determine how much disturbance a community can withstand.

CORAL REEF COMMUNITIES

Coral reefs have been present in tropical oceans for at least 60 million years, and this long history has produced the great diversity of organisms

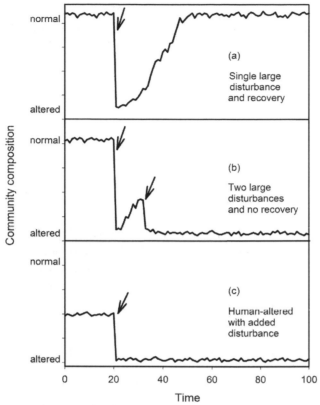

Figure 5.1 A simple illustration of the effects of disturbances (arrows) on ecological communities. (a) A community is subjected to a single large disturbance like a fire at time 20, and then recovers through a process of succession to its original state. (b) A community is subjected to two disturbances (arrows) in sequence at times 20 and 35, and the combined effects lead to a change in community composition and no recovery. (c) A community already altered by human activities like farming or forestry is then subjected to a natural disturbance, such as fire or flooding, and the combination of stresses changes community composition and prevents it from recovering in a short time. "Normal" is used here as a shorthand for "previous community state." (Modified from Paine et al. 1998.)

that are present on reefs today. Coral reefs have long been viewed by ecologists as the classical equilibrium community living in tropical waters, constant in composition and relatively unaffected by disturbances. Our view of coral reefs now is very different because we have the benefit of long-term studies of coral reefs and of recent dramatic events affecting corals worldwide.

Coral reefs are subject to a variety of physical disturbances associated with tropical storms. On the reef surrounding Heron Island, at the southern edge of the Great Barrier Reef of Australia, Connell et al. (1997) followed changes in coral cover over a 30-year period using permanently marked areas across the reef. They measured the percentage of the area covered by corals to estimate abundance. To measure the addition of new corals on sampled areas they took sequential sets of photographs of the same areas over several years.

Violent storms were the main source of disturbance to the Heron Island reef, and the amount of damage caused by cyclones was strongly affected by the position of the coral colonies on the reef (Figure 5.2). Five cyclones passed near to Heron Island during the 30 years of study from 1962 to 1992. Of the three study areas shown in Figure 5.2, only the protected area of the inner flat was relatively unaffected by cyclones. Virtually every cyclone caused a reduction in coral cover in the exposed pools. The 1972 cyclone completely removed coral cover on the exposed crest, the most severe disturbance observed. Recovery on the exposed crest was slow for the next 25 years. Gradual declines in coral cover on the protected sites was caused by increasing exposure to air as the corals grew upward over the 30 years of study.

Coral reefs have been under stress around the world and the picture that emerges from this work on the Great Barrier Reef is of a coral community that changes continually because of disturbance caused by tropical cyclones and internal processes of growth and recruitment. The coral community is not constant at the spatial scale of the reef because the frequency of disturbance is greater than the rate of recovery.

Like natural disturbances to coral reefs from cyclones, human disturbances also cause coral reef declines. One of the worst problems occurs from blasting, a crude form of illegal fishing on reefs. The detonation of homemade bombs not only kills fish (which are collected) but also shatters the coral skeletons, creating expanses of unstable coral rubble that reduces survival of coral recruits because the rubble moves around due to wave action. Fox and Caldwell (2006) showed that single blast sites recovered their hard coral cover within 5–10 years (Figure 5.3). However, if many blasts are concentrated in a larger area, a rubble field is created that prevents recovery for a period from several decades to centuries, even if reefs are protected from further blasting.

Marine protected areas are one way to prevent the continual erosion of coral reefs around the world. Selig and Bruno (2010) showed that marine

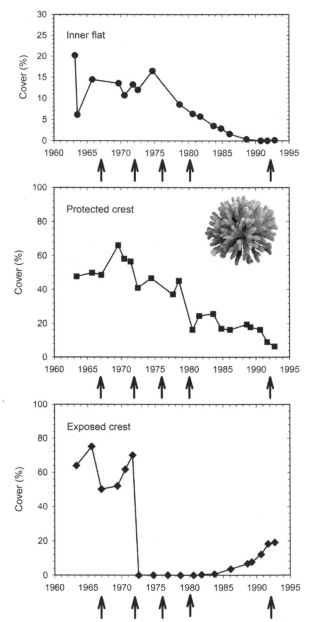

Figure 5.2 Coral cover on three areas of reefs surrounding Heron Island at the southern edge of the Great Barrier Reef, Australia. Years with tropical cyclones are indicated by arrows. Permanent quadrats were measured in these shallow water sites from 1963 to 1992. Damage from cyclones was highly variable depending on how much the site was protected by the island. (From Connell et al. 1997.)

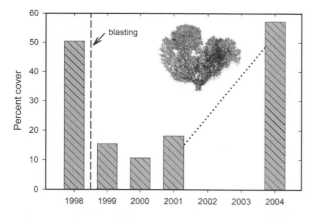

Figure 5.3 Recovery of coral from a single blast from illegal fishing. Immediately after the blast area covered by live corals is reduced to about one-fourth of its previous size. The corals recover over a period of 5 years. (Data from Fox and Caldwell 2006.)

protected areas not only increased the local fish populations but also protected corals from human disturbances associated with fishing. They also showed that the longer the protection of a coral reef, the greater the improvement in coral cover. Closing marine protected areas to protect corals need not eliminate fishing, as long as regulations are in force to limit fishing intensity and to prevent illegal practices like blasting for fish capture.

Coral reefs cannot be protected from physical disturbances from typhoons or tidal waves but they can be protected from human destruction. A variety of pollution and sedimentation problems clearly affect reefs, as well as outbreaks of disease and coral bleaching from high sea water temperatures. All of these stressors must be managed as much as possible to allow coral reefs to bounce back from natural disturbances (Hughes et al. 2010, Graham et al. 2011).

ARID ZONE VEGETATION RECOVERY

Arid zone vegetation has been disturbed by humans for many years. In the warm deserts of the southwestern United States, many studies of plant recovery after land clearing for pipelines, roads, mining, and towns now abandoned have provided a clear picture of how plant communities in this arid zone can recover after disturbances. Abella (2010) summarized the many studies of recovery in the Mohave and Sonoran Desert regions of the USA. Recovery in arid lands is entirely from natural processes since there is rarely money for land management or restoration plantings. Only discrete disturbances were studied in this review by Abella (2010); thus, grazing impacts are not included because they are a continuous, long-term disturbance.

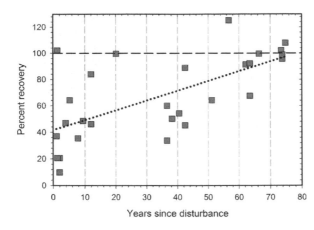

Figure 5.4 Plant cover in the Mojave and Sonoran Deserts of the American Southwest in relation to time since disturbance. Plant cover is expressed as the percent of levels found on undisturbed areas. Horizontal dashed line indicates complete recovery. (Data from Abella 2010.)

The simplest metric for plant recovery after disturbance is time. Figure 5.4 shows the rate of recovery of perennial plant vegetation cover with reference to adjacent undisturbed sites. In this arid region complete recovery (if there is no additional disturbance) would take about 70 years. In these arid communities there was no change in the number of plant species present over time, but in general there was a succession from short-lived species to longer-lived ones. Disturbance types affected the rate of recovery of vegetation. After a fire, vegetation recovery was more rapid than after other disturbances like land clearing for roads or town sites. But given a long time frame, recovery of disturbed sites to that of the original community was very good.

PLANT SUCCESSION ON VOLCANOS
Volcanic eruptions typically produce an area that is nearly completely sterilized of all life, and such an area is a good place to study recovery from severe disturbances. Mount St. Helens in southwest Washington State erupted catastrophically on May 18, 1980, destroying the mature forest that occupied its slopes. About 400 m was blown off the cone of this volcano, and the blast from the eruption devastated a wide arc extending some 18 kilometers north of the crater. The eruption produced a landscape with low nutrient availability, intense drought, and frequent surface erosion, a great variety of conditions for vegetative recolonization.

Colonization of habitats above treeline on Mount St. Helens has been slow. Figure 5.5 shows the development of vegetation on the Pumice Plains, 4.5 kilometers northeast of the lava dome. There were no surviving plants in this site in 1980, and only 4 species by 1989. But species con-

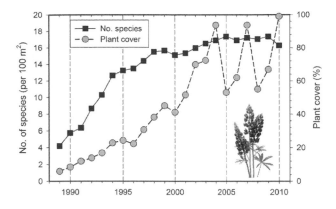

Figure 5.5
Development of the plant community on the Pumice Plains of Mount St Helens after the volcano erupted in 1980 and plant cover was obliterated. The colonizing plant community was dominated by lupine. Summer drought reduced plant cover in 4 years. (Data from del Moral et al. 2012.)

tinued to colonize the area, and by 2000 the species list reached a plateau of 15 species. Only a few additional species invaded after 2000, and the peak of 17 species was reached in 2005. Plant cover has increased slowly so that complete plant cover was reached on this site only 25–30 years after the eruption (del Moral et al. 2012).

Early primary succession on volcanic substrates rarely produces plant densities sufficient to inhibit the colonization of other species. Neither space nor light are limiting resources for plants in this environment. So-called nurse plants facilitate the establishment of other species. Lupines (*Lupinus lepidus*) have heavy seeds and are poorly dispersed, but they eventually become locally common on mudflows and broken volcanic rock surfaces (del Moral and Wood 1993). Before lupines get very widespread, wind-dispersed plants such as *Aster ledophyllus* and *Epilobium angustifolium* become established in lupine clumps and survive better in the shelter of these nurse plants. Individual lupine plants die after four or five years, but because they fix nitrogen they contribute on a local scale to increased soil-nitrogen levels.

Chance events strongly affected primary succession on Mount St. Helens. Biological mechanisms are initially very weak in the severe environments produced by volcanic flows. The ability to become established in these severe environments is directly related to seed size, but dispersal ability is inversely related to seed size. Consequently, subalpine areas on Mount St. Helens received many wind-blown seeds, but almost none of these small seeds germinated and achieved colonization under the stressful conditions. When plants with large seeds colonize by chance, they become a focus of further community development. If a single plant survives in the devastated landscape, it quickly becomes a locus for seed dispersal to adjacent areas, so positive feedback occurs during the early years of succession. Primary succession on Mount St. Helens has been very slow because of erosion, low-nutrient soils, chronic drought stress, and limited dispersal of larger seeds to areas distant from undisturbed vegetation.

Mount St. Helens provides a graphic example of plant succession after extreme disturbance. Measuring the speed of change on the mudflow areas allows us to estimate that it will require more than 100 years for the landscape on Mount St. Helens to return to a mature forest plant community. Understanding succession also requires understanding the mechanisms that drive changes in communities, and one focus has been on the effects that early successional species have on later successional species. Early species can help, hinder, or not affect the establishment of later species. Competition between individual plants for resources such as water, light, or nitrogen may drive succession. On Mount St. Helens we can see these processes occurring slowly, which enables us to understand them more easily. Understanding how naturally disturbed landscapes renew themselves can help us to understand how human-disturbed landscapes might respond to disturbances.

The volcanic eruption of Krakatau in islands off the east end of Java was one of the largest eruptions in modern times. A major eruption occurred about 60,000 years ago, and then the islands gradually rebuilt. The volcano remained dormant until May 20, 1883, when new eruptions started. After eruptions of increasing violence, a catastrophic explosion of August 27, 1883, repeated the prehistoric collapse and left three islands completely sterilized and covered in pumice and 60–80 m of ash. Rakata is the largest island remaining after the eruption, and its history of revegetation has been studied since 1886 (Whittaker et al. 1989).

The ash layer was nutrient-rich, and a good soil for plant growth except for the initial lack of organic matter. The Krakatau Islands have been little disturbed by humans, and are now a protected area. The colonization of these islands by plants is thus mostly natural by wind-blown seeds, seawater flotation, fruit bats, and birds. In October 1883 and May 1884 there were no plants on Rakata. By 1886 there were 25 species of algae and plants present, including 6 species of blue-green algae and several ferns The continuing increase of species over 100 years is due to plant succession from the early pioneer species to grasslands and finally to forest. Over the past 100 years seed plants have gradually come to dominate the flora (over 300 species). Spore-producing plants include many ferns, horsetails, and club mosses and comprise 99 species. By 1983 about 400 species of plants had colonized this island (Whittaker et al. 1989).

For most of these colonizing species the mechanism of dispersal can be described. For the seed plants over the entire 1883–1983 period the presumed method of arrival were: wind, 99 species; sea flotation, 103 species; animal-carried, 123 species; and human introductions, 32 species. The animal-dispersed plants were typically carried by fruit bats and birds.

Volcanos provide a model system for studying plant and animal succession over time following disturbance. The key variables affecting the rate of colonization are the distance to the nearest undisturbed source area, the type of soil deposited by the eruption, and the climate (rainfall and temperature). Without plants, no animals can subsist, and hence most volcano studies have concentrated on plants.

The most recent volcano eruption that has been studied is that of Mount Pinatubo on Luzon Island near Manila on June 15, 1991, the second largest volcano eruption in the 20th century. It injected large amounts of aerosols and volcanic dust into the stratosphere, second only in amount to that of Krakatua in 1883 and ten times that of Mount St. Helens. Over the following months, the aerosols formed a global layer of sulfuric acid haze, and average global temperature dropped by about 0.5°C. De Rose et al. (2011) measured the rate of recovery of 22 upland watersheds on the mountain by satellite imagery. From satellite data we derive the NDVI (Normalized Difference Vegetation Index), which measures the amount of green reflectance from any land surface. As the areas denuded by the eruption recover their vegetation, the NDVI increases; 10–16 years after the 1991 eruption, scientists were able to measure the overall recovery rate. The observations suggest that about 50 years are required for hill slopes to recover a dense vegetation cover. The use of satellite data does not provide the details of which species recovered at which rate, but it is most useful to get an overall measure of the recovery needed to stabilize mountain slopes denuded by severe disturbances like volcano eruptions.

The recovery of vegetation after volcanic eruptions illustrates well the general principle that even after severe disturbance, plant communities can return to a state close to their former composition and abundance. The key to recovery is seed and spore dispersal, and the exact vegetation path of recovery is often driven by which plants get there first.

PLANT SUCCESSION ON ABANDONED FARMLAND

Agricultural fields are often abandoned or set aside as conservation reserves, and then begin to recover with or without active management. Plant community recovery in these cases differs from that of volcanos because there is already a legacy of plants and seed in the soil on abandoned farmland, and typically an array of agricultural plants and weeds remaining on the site. Restoration ecologists wish to return these degraded sites to something like the original vegetation, and the main question is whether this is possible and how long it might take to achieve this goal. In practice the management issue is whether exotic plants ("weeds") restrict the colonization and abundance of the native species. In many present-day agricultural landscapes, native plant communities have been greatly reduced to small remnants, whereas exotic plants have become widespread.

Tognetti et al. (2010) examined trends in native and exotic species abundance over 20 years of old-field succession on abandoned farmland in the Inland Pampa, near Buenos Aires, Argentina. The farm had operated for 60 years, growing wheat, maize, sorghum, and sunflowers. Exotic species were a major component of old-field communities. They recorded a total of 149 herbaceous species, of which 40% were exotics introduced to the original pampas to improve cattle grazing. Many of the exotic species were annuals, while the vast majority of native plants were perennials. As recovery proceeded over the 20 years of study, the number of species declined at an average rate of one species lost per year (Figure 5.6a). Most of the species lost were annuals and exotics, with only a few losses of native perennials. While the number of exotic species fell over time, the dominance of the exotic species continued (Figure 5.6b). Between years 2 and 16 of the succession, two exotic species—Bermuda grass (*Cynodon dactylon*) and annual ryegrass (*Lolium multiflorum*)—dominated the site, making up 67% of all the plant cover until the last 4 years of the study.

Exotic species are rarely replaced by the native species during plant succession in abandoned farmland, and one of the important foci of restoration studies is to determine whether there are management techniques that can assist the recovery of native species. Table 5.1 summarizes some generalizations that plant ecologists have made, based on studies of plant succession, that can be useful in the restoration of ecological integrity on disturbed sites. The difficulty is that with extensive disturbances like forest fires or volcanic eruptions, the areas affected are too large to manage with limited budgets, and natural processes dictate the course of plant recovery.

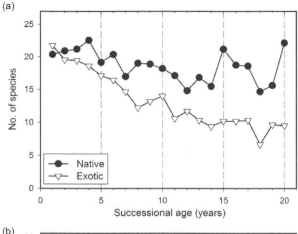

(a)

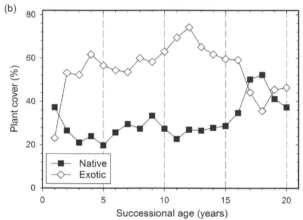

(b)

Figure 5.6 Trends in (a) total species and (b) cover of native and exotic plant species during 20 years of old-field succession in the Inland Pampa of Argentina. (From Tognetti et al. 2010.)

SMALL MAMMAL DECLINES IN NORTHERN AUSTRALIA

An underlying assumption of restoration ecology is that communities of plants and animals will recover from disturbances if there is sufficient time and if seeds, spores, or colonizers are available. A corollary of this idea is that species in protected areas like national parks should remain relatively stable and of little conservation concern. This is not always the case, and northern Australia provides a good example of a collapse of native small mammals and the possible factors affecting recovery.

Kakadu National Park is a large park (20,000 square kilometers) in the subtropics of the Northern Territory of Australia. It is a biodiversity-rich conservation reserve and is situated within a region subject to relatively little human modification. But this park has experienced a rapid and

Table 5.1 Examples of restoration tactics to address problems at crucial stages of the restoration process for disturbed sites. (From Walker and del Moral 2009.)

Restoration topic	Goal	Tactics
Establishment	Ameliorate stress	- Create safe sites to enhance survival - Install fences to trap seeds - Install perches to enhance dispersal - Fertilize appropriately - Densify stocking rates to create "nurse plant effects"
Carbon accumulation	Accelerate development	- Prepare surface (e.g., mulch) - Plant mature plants directly - Stabilize erosion - Limit grazing
Nutrient dynamics	Increase availability	- Adjust fertility by use of nitrogen-fixing species - Add carbon (sawdust) - Add phosphorus and organic matter in later stages
Life history	Enhance diversity	- Consider local species pools - Modify site for a mix of growth forms - Select species based on their weakest link (e.g., seedling survival)
Species interactions	Self-sustaining species	- Limit competition from nutrient-responsive species through planned disturbances - Include shade-tolerant species - Include nitrogen fixers

severe decline in the native mammal fauna over the last 15 years (Woinarski et al. 2011), and there is no sign of recovery. This collapse in the numbers of small mammals has occurred across most of northern Australia and is not confined to this park. Four causes for this decline and failure to recover have been suggested: increased fire frequency, increased predation by introduced cats, habitat degradation by cattle grazing, and disease.

Disease investigations have turned up nothing to suggest that it is a key factor. Fire frequency has increased in much of northern Australia in recent years, but in the investigation of the ecological effects of different burning regimes, there was no suggestion that fire frequency affected small mammal diversity or abundance (Woinarski et al. 2004). Two processes may be important—cat predation and cattle grazing. Increases in cat predation could be important, but the critical studies of cats in northern Australia are only now being started. Cats remain a significant exotic predator, and many conservation ecologists strongly believe that cats are a critical threatening process for many species in Australia (Woinarski et al. 2011). We must await critical studies on cats to determine whether they are a key limitation on recovery.

One experiment has been done on the impacts of cattle grazing on small mammal recovery. Legge et al. (2011) carried out a landscape-scale destocking experiment in the central Kimberley, northwest Australia. They measured changes in the small- to medium-sized mammal community between 2004 and 2007 following the experimental removal of cattle, horses, and donkeys from 40,000 hectares in the Mornington Wildlife Sanctuary, and compared changes in the mammal community with those in adjacent areas that still contained cattle, horses, and donkeys. Fire frequency in this area is low, so potential effects of fire frequency changes are not an important factor. The question was whether or not mammal populations would recover if large herbivores were removed from a site. Figure 5.7 shows the response to destocking this area for 3 years. Recovery was rapid in this large area within 3 years of destocking. Both the number of species and their abundances increased, and the cover of grasses and forbs on the trapping areas increased over the 3 years of the experiment. These data indicate clearly that introduced large herbivores reduce the numbers of small mammals on these sites but that other factors like cat predation may also be operating (Woinarski et al. 2011).

One interesting corollary of the decline of small mammals in northern Australia is that whatever factors are causing this drop in mammals, there

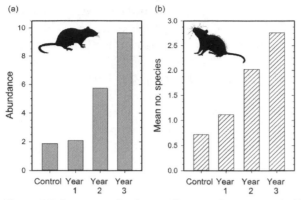

Figure 5.7 Response of native small mammals to removal of introduced herbivores at permanent quadrats within the Mornington Wildlife Sanctuary in northwestern Australia over 3 years. Control values are from normally stocked areas outside the sanctuary. (a) Mean abundance per quadrat; (b) mean number of small mammal species per quadrat. (Data from Legge et al. 2011.)

is no effect on bird numbers in the same locations (Woinarski et al. 2012). Most bird numbers in Kakadu National Park are increasing rather than declining, so that the factors responsible for the mammal collapse are not operating on all the biodiversity of these areas in northern Australia.

Two general messages come out of these studies in northern Australia. First, disturbances in communities may affect some species in the community but not others. This makes it imperative to find the processes by which disturbances affect different species. Second, once a threatening process is taken away, communities may recover rapidly.

CONCLUSIONS

Ecological communities contain many species of plants, animals, and microbes, and they are continually subject to disturbances caused by both physical and biological factors. Fires, natural disasters, drought, temperature changes, introduced species, and seasonal events all influence changes in abundance of the species in the community The introduction of human disturbances in addition to these natural disturbances has produced questions about how much disturbance a community can tolerate before it fails to recover to a condition approximately similar to its starting state. The job of the community ecologist is to map these responses to disturbance and to identify the paths to recovery and the time frame required.

These studies of recovery lead us into two different directions. We need to ask what happens if the community is pushed too far and cannot recover to its original state, and this is the subject of the next chapter. The second applied aspect of disturbance ecology is the restoration of disturbed areas like toxic waste sites, overgrazed pastures, and eroded landscapes. Questions of restoration ecology will come back in the following chapters to address what are some of the most pressing problems of our age.

CHAPTER 6

COMMUNITIES CAN EXIST IN SEVERAL CONFIGURATIONS

KEY POINTS

· All biological communities are structured in food webs determined by who eats whom. Food webs are impacted by physical factors—weather, fire, and human actions.
· Regime shifts are sudden strong reorganizations of biological communities that have large effects on the food web. They are tipping points that are more easily recognized in marine and freshwater communities but also occur in terrestrial systems.
· The vital issue is whether community changes can be reversed. There are enough examples now of cases where the regime shift is a permanent change in which species are lost from a community and new species enter. These kinds of regime shifts have many consequences that we cannot ignore.

We tend to assume in our relatively short lives that the ecological communities we see are permanent, so that when we walk through the red-woods of California or the beach grasses of North Carolina we assume that nothing will change in our lifetime. But ecological communities, as we saw in the last chapter, are often hit by sudden disturbances; what happens then? Ecological communities, much like our human communities, can often rebound from a single small disturbance and return to their starting configuration. But if a disturbance is sufficiently drastic, or if the disturbing forces become permanent, the ecological community may shift to a new configuration and stay there. What happens when a biological community is so severely disturbed that it cannot recover its original configuration? If it changes to another configuration, what will it be? Can ecologists predict the consequences of large disturbances?

FOOD WEBS

To answer these questions we must first determine how a community is put together. The interactions between the species in a community is clearly seen in its food web, which is based on the simple idea of who eats

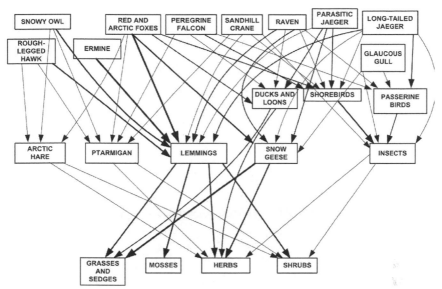

Figure 6.1 Simplified food web for the high arctic tundra site at Bylot Island, Canada. The width of the arrows indicate the relative strength of the interactions of who-eats-whom. (Data from Legagneux et al. 2012.)

whom. We classify the major organisms in a community into three categories:

- producers = green plants
- consumers = animals
  - primary consumers = herbivores (eat green plants)
  - secondary consumers = carnivores (eat herbivores)
  - tertiary consumers = carnivores (eat other carnivores)
- decomposers = fungi and bacteria that break down dead plant and animal matter

Figure 6.1 gives a simple example of a food web from arctic Canada. Food webs are organized by two major processes—by predation and by competition. Thus, for the tundra of Bylot Island in northern Canada (Figure 6.1) lemmings can compete with one another for food or nesting sites, while tundra grasses and sedges compete for space in which to grow and for nutrients and water in the soil. Predation and competition always interact; what happens in the community is an overall summation of hundreds of species interacting through predation and competition. Perturbations to a community act through the linkages of the food web.

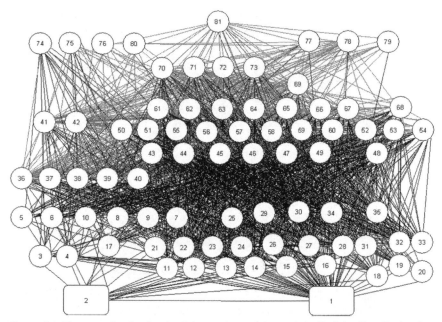

Figure 6.2 The complex food web of the marine continental shelf of the New England states and eastern Canada. Phytoplankton (2) are at the base of the open ocean food web, and a host of plankton (3–4) and shrimp (8–12) are the herbivores at the next feeding level, blending into the small fishes (36–40) as predators on the herbivores and up to the higher predators that eat the smaller predators like seals (76) and sharks (78). Humans (81) occupy the top of this food web. (From Link 2002.)

The tundra food web is relatively simple because this is a high arctic community, but it illustrates two things about food web construction. First, some species are grouped together. For example, the grasses and sedges compartment contains about 15 species, and the passerine bird and shorebird compartments contain about 12–15 species. If the purpose of the study is to concentrate for example on shorebirds, these compartments would have to be subdivided and measured as separate units. Second, some links are stronger than others (shown by the thick lines). These links may be the critical ones affecting the composition of the community and its functioning.

Food webs can grow very complex in polar, temperate, and tropical communities, even if only the major species are identified. Figure 6.2 shows the food web of the continental shelf of New England and eastern Canada, and the central position of invertebrates like shrimp and small fishes in this complex food web. Humans are at the top level of this marine food web and can potentially compete with seals and sharks for food. If strong pressure is exerted on any species in this web, it can potentially reverberate throughout the web because of the tangle of feeding relationships.

The food web shown in Figure 6.2 is complex but is still highly aggregated at the level of the invertebrates toward the base of the food web. And the bacteria and viruses that live in this marine environment are not included. This complexity shows clearly that ecologists cannot study all the linkages in a complex food web but must subdivide the web, restricting the sub-web to those species that interact directly with the species of interest.

Knowing the food web gives ecologists some insights into how the system is constructed but not necessarily how it works. This is because there is one large element missing from food webs—the impact of the environment on the whole community. In this era of global warming the environmental effects of increased temperatures or reduced rainfall can have important ramifications for community structure and species interactions. The clearest insights come from major environmental shifts associated with climate that lead to such large changes in communities they are called "regime shifts."

REGIME SHIFTS IN OCEANIC PRODUCTION

Regime shifts are sudden shifts in ecosystems to another state in which the abundances of many species in the community change and some species might disappear. The most dramatic ones have occurred in the oceans. Figure 6.3 shows a regime change in the North Pacific. Changing oceanic currents move warm and cool water north or south, showing up as a rapid shift in ecosystem productivity.

The explanation for the regime shift in the North Pacific has been climatic forcing as it affects seasonal phytoplankton cycles. As ocean currents change, the warm water of some currents is replaced by the colder water of other currents. This change in water temperature shifts the length of the growing season for phytoplankton and thus for zooplankton that feed on them (Figure 6.4). Before the mid-1970s, winters were relatively warm and summers were moderately warm, and the peak abundance of the spring and spring–summer copepod communities occurred in April and July. After the 1976 regime shift to colder winter temperatures, the timing of the peak abundance of plankton was delayed by 1 month in the spring community, in response to the shift in the timing of the spring bloom. The summertime conditions shift back and forth between relatively warm and cold water, as did the winter and spring conditions. The

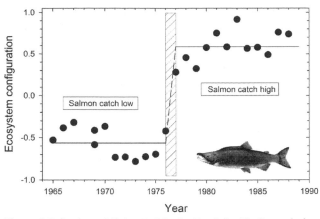

Figure 6.3 Regime shift in 1976 in the North Pacific Ocean indexed by a suite of 100 environmental and fish time series combined into an index that measures the overall organization of the ecosystem. Around 1990 the regime shifted back to the state found in the late 1960s and early 1970s. In the first state, the Pacific coast salmon fishery caught relatively few fish, and in the second state the salmon catch increased dramatically. Ocean temperature is the main driver of these shifts. (Data from Hare and Mantua 2000.)

warmer surface water permitted more plankton growth and this gradual change reached a tipping point because many invertebrates and fishes thrive in warmer surface waters. But summertime warming becomes self-limiting because it produces strong water stratification to a level that reduces nutrient availability for phytoplankton growth. Poor phytoplankton availability then reduces the July abundance of the spring and summer community of fishes and invertebrates. These regime shifts are driven by temperature and currents and are a gradual process that reaches a tipping point for some species like salmon that suffer food shortage and lower survival. It happens over a time scale of a decade in the oceans, while a similar regime shift is now occurring on land via climatic warming, but with a time scale of perhaps a century.

Regime changes show that some ecosystems can undergo abrupt transformation in response to relatively small environmental changes. If ecologists can locate these tipping points, it can assist in managing for biodiversity conservation, particularly in the face of climate change.

One mechanism likely to induce widespread regime shifts in polar ecosystems is the loss of sea ice due to global warming. The angle of the sun varies dramatically with the seasons in the polar regions, and the amount

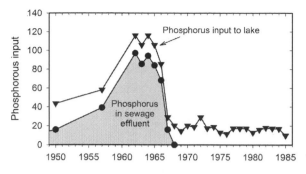

Phosphorus input to lake

Phosphorous input

Phosphorus in sewage effluent

1950 1955 1960 1965 1970 1975 1980 1985

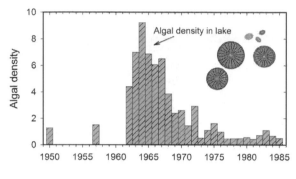

Algal density

Algal density in lake

1950 1955 1960 1965 1970 1975 1980 1985

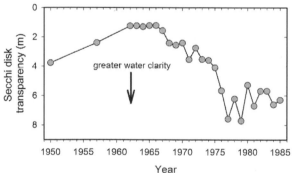

Secchi disk transparency (m)

greater water clarity

1950 1955 1960 1965 1970 1975 1980 1985

Year

Figure 6.4 Recovery of Lake Washington in Seattle, 1950–1984. Treated sewage flowed into the lake in increasing amounts during the 1950s. Sewage was diverted from the lake gradually from 1963 to 1968. The phosphorus content of the lake decreased rapidly after sewage was diverted. Blue-green algal density dropped in parallel with phosphorus levels because phosphate is the nutrient that limits blue-green algal growth in this freshwater lake. Zooplankton eat small green algae (but not blue-green algae), so green algal abundance also fell because of grazing by zooplankton and the lake water became clearer. (Data courtesy of W. T. Edmondson.)

of light entering the water column depends critically on how long sea ice a meter or more thick covers the surface of the water (Clark et al. 2013). If the ice-free period is 100 days long, four times as much light will enter the water column as in a 50-day ice-free period. Light is a key driver of community structure on shallow seabeds, regulating interactions between large algae and invertebrates. Algae need light for photosynthesis, and invertebrates do best in dark areas of the seabed and so are outcompeted for space by algae in well-lit areas. Significant increases in light could switch shallow water communities from invertebrate-dominated to algae-dominated states.

Clark et al. (2013) studied shallow water sites near Casey Station in East Antarctica to measure the relationship between the ice-free season and community structure of algae and invertebrates. Algae tend to replace invertebrate species when the ice-free season is longer and more light is available, and invertebrates dominate when the season is short. Climatic warming will push this community from the current invertebrate-dominated situation to an algae-dominant one when the ice-free season increases. Approximately one-third of the species in the shallow water community in this part of Antarctica could be locally extinct once the coast is regularly ice-free for more than half the year. Invertebrates at risk of local extinction include filter-feeding sponges, bryozoans, tunicates, and polychaetes, all of which perform vital ecosystem functions such as water filtration and nutrient recycling. These fauna are critical components of the coastal Antarctic food web.

REGIME SHIFTS IN FRESHWATER LAKES

Many freshwater lakes have changed dramatically as a result of nutrient additions from human activities. Lake Washington is a large, formerly unproductive, lake in Seattle, Washington, that was used for sewage disposal until the late 1960s. By 1955 it was clear that the sewage input was destroying the clear-water lake, and a plan to divert sewage from the lake was voted into action. More and more sewage was diverted to the ocean from 1963 through 1968, and almost all was diverted from March 1967 onward. Thus the recent history of Lake Washington consists of a pulse of nutrient additions, followed by a complete diversion.

Since the diversion of sewage began in 1963, ecologists from the University of Washington have recorded the changes in Lake Washington. Figure 6.4 shows the rapid drop in phosphorus in the surface waters and the closely associated drop in the standing crop of phytoplankton. Phosphorus is a limiting nutrient to phytoplankton growth in most freshwater lakes. The water of the lake has become noticeably clearer after the sewage diversion. Phosphorus is tied up in the lake sediments but is released back into the water column rather slowly.

The Lake Washington experiment suggests that detrimental changes in lakes may be *stopped and reversed* if the input of nutrients can be stopped. The restoration of Lake Washington shows that this community can exist as a clear water lake with low nutrient inputs or as a lake with high nutrient inputs stimulating algal growth and green-colored water. The plankton and fish community of Lake Washington changed greatly as the nutrient input dictated the conditions for algal growth.

One of the changes that often accompany nutrient pollution in freshwater lakes is that blue-green algae tend to replace green algae. Blue-green algae are often called "nuisance algae" because they become extremely abundant when nutrients are plentiful and form floating scums on lakes. Blue-green algae become dominant for several reasons. They are not heavily grazed by zooplankton or fish, which prefer to eat other algae. Some species of blue-green algae also produce secondary chemicals that are toxic to zooplankton as well as humans. Blue-green algae are also poorly digested by many herbivores, so they are low-quality food for them. As more and more phosphorus is continually loaded into a lake from fertilizers or from sewage, the nitrogen-to-phosphorus ratio falls and nitrogen becomes a limiting factor. Many blue-green algae can fix nitrogen, so that when nitrogen is limiting, they are at an advantage over green algae that cannot fix nitrogen. The phytoplankton community in many temperate freshwater lakes may have two broad configurations at which it can exist, one with low nutrient levels (dominated by green algae) and one with high nutrient levels (dominated by blue-green algae).

Regime shifts in freshwater lakes may be caused by changes to the predator population as well as by shifts in the nutrient inputs. A simple experiment to demonstrate such a transition was done by Pace et al. (2013) from 2008 to 2011 on two lakes in Michigan. Paul Lake is dominated by largemouth bass (*Micropterus salmoides*) and also supports a small population of pumpkinseed sunfish (*Lepomis gibbosus*). These fishes feed mainly on smaller fishes in the lake, and this lake served as the unmanipulated control for the experiment. In 2008, Peter Lake, the manipulated lake, was dominated by a variety of small fishes that consumed plankton and had only a few largemouth bass in it. From 2008 to 2011 largemouth bass were added to Peter Lake to determine if these predators could shift Peter Lake into a configuration like that of the control. Figure 6.5 shows the results of this 4-year experiment. The abundance of largemouth bass increased with the additions in the experimental lake, and at the same time the abundance of the small plankton-eating fish dropped abruptly. Predation by largemouth bass was sufficient to shift this lake into another configuration where the dominant biomass was of predatory fish.

The addition of predatory fishes reduced the abundance of their prey

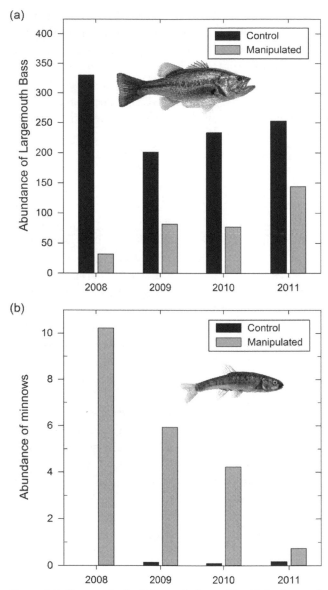

Figure 6.5 Changes in the fish populations of Paul Lake (control) and Peter Lake (manipulated) from 2008–2011. The increase in the predatory largemouth bass in Peter Lake was reflected in the collapse of the small prey minnows eaten by bass. (Data from Pace et al. 2013.)

(minnows of several species) and the reduction of minnows in turn allowed the plankton community to shift toward the dominance of larger-bodied plankton species. The shift to larger species of plankton changed the phytoplankton dynamics, so that the changes in the top predator in these lakes has effects that cascade down the food web from the predator to the plant trophic level. Regime change can come from either end of the food web, from the plants (bottom-up) or from the predators (top-down).

A SIMPLE GRAPHICAL MODEL OF REGIME SHIFTS

The dynamics of regime shifts in plant and animal communities can occur in three different ways, illustrated in Figure 6.6.

The two key points to applying these ideas to community changes are to know what the controlling variable is and to define carefully the community trait. Often the community traits measured are the number of species in the community and the abundance of the species. In an extreme case some species in a community may go locally extinct if a community shifts from one state to another; the key information then needed is whether the missing species can return via dispersal from adjacent sites. Terrestrial plant species are often more resilient than animal species because they have a seed bank in the soil and can return when favorable conditions occur.

REGIME SHIFTS IN TERRESTRIAL COMMUNITIES

Many changes occur in plant communities because of fire or other strong disturbances; there is considerable discussion about which of the three models in Figure 6.6 best describes the change. A key question is whether changes in plant communities are reversible or not on a human time scale. One example to illustrate this issue is the conversion of open savannas to closed woodland.

Savannas are mixed systems characterized by a tree canopy in an open and continuous grass layer. Because tree cover in savannas varies from very sparse to quite dense, the extremes (open grasslands and closed woodlands) have sometimes been interpreted as alternative stable states. The key defining feature of a savanna is the dominance of grasses in the herbaceous layer. Grasses in higher-rainfall savannas support frequent fires, which may select for fire-tolerant trees and forbs. If fire is absent the community changes to thickets and ultimately forests, where trees shade out grasses and fire-sensitive species thrive. A combination of savannas and forests could be considered an alternative stable state to savannas alone.

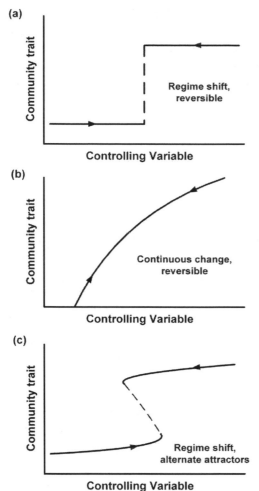

(a)

Community trait

→

Regime shift,
reversible

→

Controlling Variable

(b)

Community trait

←

Continuous change,
reversible

↑

Controlling Variable

(c)

Community trait

←

Regime shift,
alternate attractors

→

Controlling Variable

Figure 6.6 Three possible modes of change in alternative community states. Any community metric like the number of plant species can change abruptly (a) as some controlling variable like temperature or rainfall or herbivore abundance changes, and this could be reversible. Many human landscape alterations fit this mode of change. Alternatively, the change may be gradual (b) and could be reversed; no tipping point would be visible. A third possibility (c) with a clear tipping point is a regime shift that puts the community in an alternate stable state that may be more difficult or impossible to reverse.

In southern Africa, woody encroachment in savannas is widespread and there are two alternative processes for the changes from open savannah to woody thickets (Figure 6.7). Parr et al. (2012) analyzed the conversion of savannah to woody thickets in the Hluhluwe Game Reserve, South Africa. Aerial photographs of the park going back to 1937 were paired with satellite images from the 1980s and on-ground surveys in 2007–2009 to evaluate which model of change was most likely for this ecosystem.

Parr et al. (2012) found that the changes in these South African savannas were not just an increase in woody cover of savanna species, but a

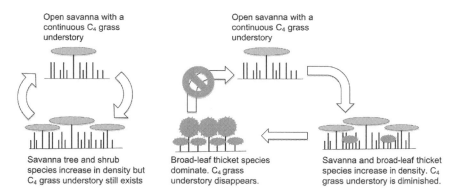

Open savanna with a continuous C_4 grass understory

Open savanna with a continuous C_4 grass understory

Savanna tree and shrub species increase in density but C_4 grass understory still exists

Broad-leaf thicket species dominate. C_4 grass understory disappears.

Savanna and broad-leaf thicket species increase in density. C_4 grass understory is diminished.

(a) Savanna thickening
Part of a cyclical process. Encroached savanna ecologically similar to open savanna, and able to revert back to open savanna.

(b) Thicket expansion
Thicket expands into savanna, eventually causing a switch to forest. Not part of a cyclical process, and unlikely to switch back to open savanna.

Figure 6.7 Schematic diagram showing two distinct woody thickening processes. (a) Savanna thickening occurs when densities of savannah tree and shrub species increase and is a cyclical reversible process. (b) Thicket expansion is a shift in composition with increased densities of broad-leaved, shade-tolerant thicket and forest species, along with the loss of grasses in the understory. Thicket expansion shades out the grass layer, and may be difficult to reverse because grasses carry fire; it may lead to a switch to thicket and even eventually forest so that grasses would be lost to the plant community. (From Parr et al. 2012.)

wholesale change in habitat (structurally and compositionally), representing a complete switch in the plant community from open grassy savanna to closed thicket (Figure 6.7b). The thicket habitat became a mix of shade-tolerant tall herbs, woody shrubs, and scattered patches of shade-tolerant grasses. The alternative states to savanna plant communities are savanna, thicket, and ultimately forest, where there is a switch in composition from grasses and forbs to woody species

The flammability of savannah (high) and thickets (low) is an important difference in these alternative communities, and this is a key element is making it difficult to reverse the shift from savannah to thickets, the equivalent of a regime shift in a terrestrial ecosystem. If fire is rare in thickets because grasses are not available as fuel, the change back from a thicket to an open savanna may be difficult without active management with controlled burns.

The ant fauna of savannas and thickets are markedly different. More ant species in thickets are predatory forest ants, a part of a change in trophic structure within this important component of the community. This shift in trophic structure is linked to the development of a leaf litter layer in the thickets. Leaf litter is a key microhabitat for invertebrates and provides critical habitat and resources not found in the open savannas. Many of the ants live and forage on rotten wood and are predators in leaf litter, preying on spiders' eggs and small insects. Savannas lack a litter layer because fires are too frequent. Birds, small mammals, and reptiles also respond to the habitat change between savannah and thickets because of the additional cover provided by the taller vegetation in thickets.

The regime shift from savannah to thickets and perhaps ultimately to woodland is of conservation concern because it appears to be a community change that is very difficult to reverse. Woody thickets are developing in savannas across much of southern Africa (Parr et al. 2012), and this regime shift has conservation consequences for biodiversity (species losses), as well as economic consequences because of the attraction of savannas and large mammals to ecotourism.

A second regime shift in Africa is working in the opposite direction from that just described for savannas. Woodlands and grasslands of east Africa may represent multiple stable states of a grazing system with two possible communities. Woodlands in parks and reserves over much of the savanna areas of east Africa have declined in the past 30 years, and been replaced with grasslands, the opposite of the result we have just described for southern Africa. Elephants are the key biological player in this change, and fire is the key physical process. Three hypotheses have been proposed to explain the woodland decline:

- Human-induced fires eliminate woodland. There is one stable state, and if fires could be reduced woodlands would return (Figure 6.6a).
- Elephants eliminate woodland by destructive browsing, and the resulting grassland is maintained by fire. There are two stable states. If fires were eliminated, woodlands would return to their former abundance, even if elephants remained (Figure 6.6b).
- Fire eliminates woodlands, and elephants hold tree regeneration in check by eating small trees so the woodlands can never return unless the elephant population is reduced by poachers or active management. Eliminating fire will not cause woodlands to return because elephants eat small trees, and there are two stable states (Figure 6.6c).

The available evidence supports the third hypothesis. During the 1960s, fire burned on average 62% of the Serengeti each year, and even with no elephants or other browsing or grazing mammals tree recruitment would be too low to sustain woodlands. Elephant and wildebeest numbers increased in the parks and reserves by the 1980s. Wildebeest grazed much of the grass each dry season so the fuel load in the 1980s was reduced and fires burned only 5% of the area each year. Elephant browsing in the 1980s was severe and by itself capable of preventing woodlands from dominating the landscape in the absence of fire. If elephants and wildebeest are reduced by poaching in the future, grass growth will increase and subsequently fires will increase to prevent woodland regeneration. At the present time the Serengeti-Mara ecosystem in east Africa seems to be locked into a grassland state and the woodlands will not return.

Another example of a striking change in plant communities comes from the eastern United States, where white-tailed deer have increased greatly in recent years. By their browsing, deer are creating alternate woody plant communities. Between 1890 and 1920 much of the hardwood forests in Pennsylvania were clear cut. These stands contained valuable trees like white ash, sugar maple, red maple, and black cherry. Deer populations increased rapidly in the regenerating stands that produced much browse for deer food. At the same time, predators like wolves were removed from the system. As a result, deer numbers skyrocketed, and deer are now considered overabundant. Deer browsing reduces hardwood tree regeneration, particularly of valuable timber trees. With sufficient browsing, the seed bank of these hardwoods becomes exhausted within 3–4 years, and without a seed bank no regeneration is possible. Ferns and grasses then invade the forest floor and suppress regeneration of desirable hardwoods completely. The result is a community of trees dominated by black cherry and other tree species less preferred by foresters for timber and less preferred for browse by deer, an alternate tree community that may be stable on the time frame of 300 years or more, even if the deer are somehow taken away. This is a good example of a community change caused by human actions of changing the landscape and removing a top predator that was keeping deer numbers low (Figure 6.6a). The results most noticed by humans in eastern North America are deer-automobile collisions that cause significant economic losses, as well as injuries to people traveling in cars and the deaths of many deer along roads.

CONCLUSIONS

Most humans prefer stability to change, yet natural communities are continually changing and can rarely be described as a stable system of plants and animals. Those changes that we see can come from physical disturbances like hurricanes or from biological changes, and unless we understand more what the impacts are of disturbances, we will not know how or whether we can reverse their undesirable effects.

Regime change is the most critical of all community changes because it comes rapidly after a tipping point is reached and it is sometimes not reversible without high economic costs. Aquatic tipping points in the marine environment are often driven by changes in water temperature and now (with CO_2 rising) with changes in acidity. Other tipping points can be driven by changes in predator abundance. When top predators like wolves and lions are exterminated in a region, there can be a strong realignment of the other predators, herbivores, and plants in the community. Other tipping points can come from disease epidemics, and the take-home message is relatively simple: try to understand the structure of the community before it is disturbed by human or natural forces, and always be prepared for surprises. Not all community and ecosystem changes can be reversed. The law of unintended consequences can have serious costs for humanity and for natural ecosystems.

CHAPTER 7

KEYSTONE SPECIES MAY BE ESSENTIAL TO THE FUNCTIONING OF BIOLOGICAL COMMUNITIES

KEY POINTS

- Keystone species are species of relatively low abundance that have a very large impact on community structure and function if they are removed from the system.
- Many keystone species are top predators, particularly in aquatic communities, but large herbivores can also have keystone effects on terrestrial communities.
- The retention of keystone species in biological communities is an important conservation objective, and large carnivore predators that are keystone species are threatened in many ecosystems by human activities.

Biological communities contain thousands of animal and plant species, and ecologists try to determine how these communities function by analyzing food webs, as we saw in the last chapter. But we can also ask whether all these species are important to community function. A species is important in this sense if its removal causes the community to change by lost production or by gaining or losing additional species. We thus ask how replaceable a species is, and how much its loss reverberates to other species in a community. This question is critical for conservation because in its extreme form it is asking the biological consequences of extinctions, and the lists of endangered species grow yearly with concerns about species that may be driven extinct.

One measure of importance is abundance, which is typically measured as numbers for animals and biomass for plants. We can begin with the simple idea that species that are most abundant, which we call *dominant species*, are the key players in community dynamics. Dominant species are determined separately for each trophic level. For example, the sugar maple is the dominant plant species in part of the deciduous forests in eastern North America, and, by its abundance, determines in part the physical conditions of the forest community. Buffalo grass is a dominant perennial in the short-grass prairie of western Kansas. Brown lemmings are the dominant herbivores on the Arctic Coastal Plain of northern Alaska, and wolves are the dominant predators in the boreal forests of Canada and Alaska and now in Yellowstone National Park.

There are three general strategies that species use to achieve dominance. First, *be quick*. A species that can find new habitats quickly, and increase in numbers quickly, can sometimes be a dominant because it got there first and achieved high abundance before any competitors arrived. Only a few species can become dominant in this way, and their dominance usually is short-lived. We call many of these species *weeds*. Second, *specialize*. A species that becomes a specialist on a resource that is very common and widely distributed can itself become common and an ecological dominant. A species that eats buffalo grass can become very common in the short-grass prairie of western Kansas. Or third, do the opposite and *generalize*. A species that can use a great variety of foods or other resources can attain numerical or biomass superiority, although it will be faced with stiff competition from other species using the same resources. Generalists are dominants only if they have high competitive ability, and many dominant species fit this third description.

Since dominant species are so numerous or contain so much biomass, we might expect them to be very important to the community. But what happens when you remove a dominant species from a community? As far as we can tell, not much; this is very surprising to ecologists. The most spectacular case is that of the American chestnut (MacDonald 2003), which disappeared from the eastern deciduous forest with no indication of damage to ecosystem functioning in the forest community. Dominant species can be removed from a community because they are involved in strong competition with other species in the same trophic level. When chestnuts died, their place in the forest was taken by oaks, hickories, red maples, and poplars—chestnut trees were completely replaceable in the forest tree community (Good 1968). Any herbivores that were completely dependent on chestnut would also have disappeared when the chestnuts died, but such specialization seems to be rare, at least in temperate communities. Dominant species are not essential to a community, contrary to our intuitive feeling that size or abundance should signify importance.

KEYSTONE SPECIES IN THE ROCKY INTERTIDAL

The species that turn out to be essential to maintaining community structure are sometimes the unexpected ones. They are called *keystone*

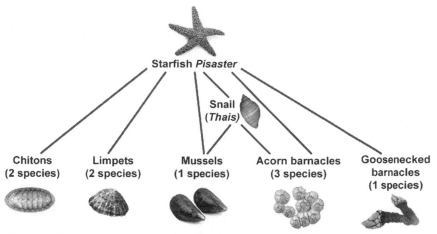

Figure 7.1 The large-invertebrate food web of the rocky intertidal zone of the Pacific coast of Canada and the northwestern USA. The starfish is the top predator in this intertidal zone. (After Paine 1966.)

species because, like the keystone or central block in an archway, their activities greatly affect the structure of the whole community.

The starfish *Pisaster ochraceous* is a keystone species in the rocky intertidal communities along the west coast of North America. The food web of the rocky intertidal is fairly simple, if we restrict ourselves to the larger invertebrates (Figure 7.1). *Pisaster* is the most common large starfish in the intertidal zone. It weighs 1–1.5 kilograms and varies from orange to purple in color. It feeds almost exclusively on mussels if given the choice, although it will also eat barnacles, chitons, limpets, and snails. The feeding of *Pisaster* on mussels turns out to be critical for many of the species in this rocky intertidal community.

The key resource here is space, because organisms cannot survive in this wave-washed environment without a firm place of attachment. Competition for space thus is all important in this habitat. You can see this very clearly because when all the rock surface is covered with organisms, they begin to grow on top of one another, as much as their structure will allow. Not all organisms are equal in competition for space, and in the rocky intertidal zone mussels are able to monopolize space. Mussels attach themselves to the rock by a strong thread. When vacant space becomes available, mussels can colonize it in two ways. Larval mussels settle from the plankton during fall and winter. Or larger mussels can migrate as

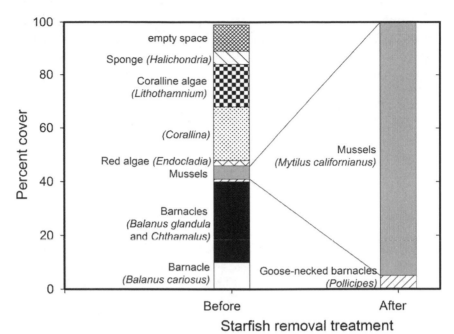

Figure 7.2 Keystone predator effect in the rocky intertidal zone of the Pacific Coast. Paine (1966) removed predatory starfish from rocky intertidal sites in Washington State and observed a collapse of the community to a near-monoculture of California mussels over 5 years. (Data from Paine 1974.)

adults by becoming detached (not voluntarily!) and being washed by waves to a new location where they reattach their thread to the substrate. When other species such as barnacles occupy a space, mussels just grow over the top of them and smother or squeeze them out.

Mussels form a tight band in the intertidal zone, and both the upper and the lower limits of this band remain stationary over the years. Larval mussels settle on rocks throughout the whole intertidal zone, but few manage to survive in the lower intertidal. What happens if one removes the major mussel predator, *Pisaster*, from these areas? Paine (1966) removed starfish from a rock face on the Washington State coast for six years, and produced a dramatic effect. Mussels began to extend their range into the lower intertidal. In 6 years they advanced downward about one meter vertically. As they advanced they took over the rock face and eliminated at least 25 species of large invertebrates and algae (Figure 7.2). The competitively dominant mussels were able to take over nearly all the space in a predator-free zone, and a near monoculture of mussels was all that remained. On undisturbed areas starfish can feed only up to a certain level on the shore because they cannot stand long periods of desiccation at low tide. Thus a band forms on the shore at the point of *Pisaster* penetration.

Starfish cannot eliminate mussels from the intertidal zone because mussels have a refuge high in the intertidal where starfish cannot feed. But there is yet another way in which mussels can elude starfish predation—by growing too large for starfish to handle. Starfish kill mussels by pulling their two shells apart, but if a mussel grows to a large size, a starfish cannot exert enough force to open it. The trick of course is to survive long enough to reach this large size, and by chance a few individuals make it. Since large mussels produce large numbers of eggs, these few individuals can contribute much to the reproductive rate of a mussel population.

The starfish *Pisaster* is thus a keystone species in the middle zone of the rocky intertidal because when it is removed, the community of organisms attaching to rocks changes dramatically in composition. Keystone species have an ecological impact disproportionate to their abundance. They can be clearly recognized by removal experiments, and they are to be looked for in communities with two features:

- some primary producer (plant) or consumer is capable of monopolizing a basic resource like space
- the resource monopolist is itself preferentially consumed or destroyed by the keystone predator or keystone herbivore.

These starfish removals have been repeated in New Zealand and in Chile, where other species of starfish eat other species of mussels. The results were identical—in the absence of starfish predators, mussels of various species tend to monopolize the rocks of the middle part of the intertidal zone. The keystone predator effect thus seems to be general to many rocky coasts of the temperate zone throughout the world.

How common are keystone species in natural communities? No one knows, yet it is clearly important from a conservation point of view that we find out. Losing a keystone species from a community may involve losing more than just one species. Keystone species may be more common in aquatic communities than in terrestrial ones.

KEYSTONE SPECIES IN THE MARINE SUBTIDAL ZONE
During the 1970s and 1980s sea otters were a keystone predator in the North Pacific. Once extremely abundant, they were reduced by the fur trade during the 19th century to near extinction by 1900. After they were

protected by international treaty, sea otters began to increase and by 1970 had recovered in most areas to near maximum densities. Sea otters feed on sea urchins, and sea urchins in turn feed largely on macroalgae (kelp). Early natural history observations showed that in areas where sea otters were abundant, sea urchins were rare, and kelp forests were well developed. Similarly, where sea otters were rare, sea urchins were common, and kelp was nonexistent. Sea otters were thus a good example of a keystone species in a marine subtidal community until about 1990. During the last 20 years sea otters have declined precipitously in large areas of western Alaska (Figure 7.3), often at rates of 25% per year. The loss of sea otters has allowed sea urchins to increase with the attendant destruction of kelp forests.

Killer whales are the suspected cause of the sea otter decline. Killer whales have begun to attack sea otters in the last 20 years because their prey base of mammals (seals, sea lions) and fishes has declined. Fishes have probably declined from human overharvesting in the North Pacific, illustrating that the interactions in food webs can propagate from top predators to basal plant species in unexpected ways. During the 1980s when seals and sea lions were abundant prey for killer whales, sea otters were the keystone species in this system. Once seals and sea lions declined during the last 20 years, killer whales became the top predator in this food web, producing the results illustrated in Figure 7.3.]

TERRESTRIAL KEYSTONE SPECIES

Terrestrial communities can also show the effects of keystone species. In east Africa the elephant is a keystone species in two ways. First, it has an impact on the vegetation community, as is described in the last chapter. Fire and elephants help preserve the grasslands and savannas of east Africa.

Second, the presence or absence of large mammals like elephants can have an impact on small mammal and snake abundance in grassland (McCauley et al. 2006); this, like the effect of killer whales on the North Pacific ecosystem, is an example of the impacts large mammals can have on their communities. The way to recognize keystone species is through experimental manipulations. The question asked in this research was: What will happen to other members of the savannah community if elephants and other large grazers like zebra are kept out of an area of grassland? Young et al. (1997) constructed a series of fences that excluded large herbivores from 4-hectare plots in Kenya in 1998. McCauley et al. (2006) sur-

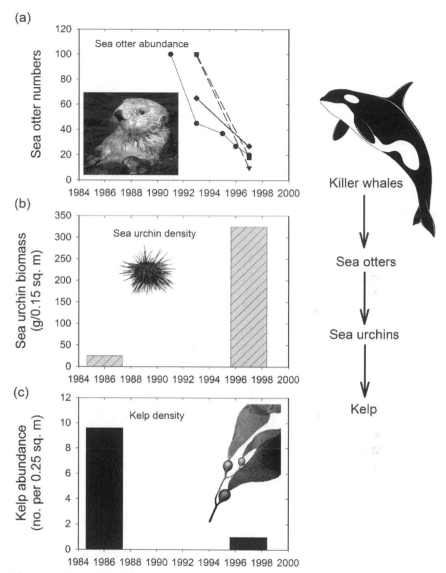

Figure 7.3 Sea otters as keystone predators in the North Pacific. Changes in sea otter abundance over time (a) at several Aleutian islands and concurrent changes in (b) sea urchin biomass, and (c) kelp density. The food chain is shown on the right. During the 1990s killer whales increased as a top predator, with the result of lower numbers of sea otters, more abundant sea urchins, and much less kelp. (Modified from Estes et al. 1998.)

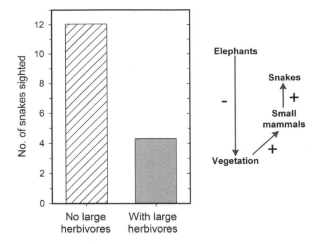

Figure 7.4 Effect of elephants on snake abundance in east Africa. The partial food web is shown on the right. Elephants depress ground vegetation by grazing and browsing. When elephants are removed, vegetation production increases and increases the abundance of small mammals, which attract more snake predators that feed on the small mammals. (Data from McCauley et al. 2006.)

veyed small mammal and snake abundance in open and fenced plots from 2002 to 2005. The most abundant small rodent in this area was the northern pouched mouse (*Saccostomus mearnsi*), and the most common snake was the olive hissing snake (*Psammophis mossambicus*), a small mammal predator. Figure 7.4 illustrates the results of this exclusion experiment.

There is no direct relationship between elephants and snakes, which move rapidly out of the way of large mammals. But there is an indirect effect of elephant abundance on snakes, so more elephants equals less snakes, but the ecological path is via vegetation and rodent abundance. Rodents increase in response to more vegetation and snakes increase in response to more rodents.

Humans are the most significant keystone predator on Earth both at the present time and in the past. One of the large episodes of extinction in the fossil record points to humans as keystone predators on large mammals and birds. About 50,000 years ago all the continents were populated with more than 150 genera of megafauna, animals larger than 44 kilograms. By 10,000 years ago more than 97% of these animals were extinct, one of the greatest extinctions of large animals known (Barnosky et al. 2004). Since many of these losses of large animals occurred near the end of the Pleistocene Ice Age, they are often grouped as the *Pleistocene extinctions*. Two characteristics of these extinctions stand out. First, they affected only large animals, and there was no unusual simultaneous loss of smaller animals or plants. Second, the losses did not occur at the same time on the different continents (Figure 7.5).

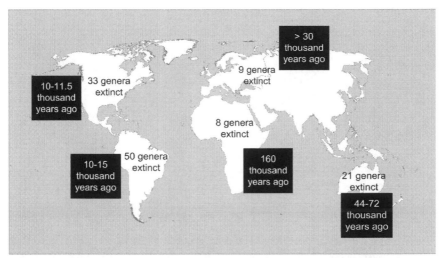

Figure 7.5 Megafauna extinctions in the late Pleistocene (from 120,000 years ago to 10,000 years ago). The numbers of large vertebrates that went extinct are given on the map of each continent. The black boxes show the timing of arrival of humans on each continent. There is a close but not perfect association between the timing of arrival of humans on each continent and the timing of the megafaunal extinctions. The recent extinction of the moas in New Zealand after the arrival of the Maori are not included here because they occurred during the last thousand years. (Data from Barnosky et al. 2004.)

The size of the lost megafauna is difficult to comprehend: giant kangaroos weighing 250 kilograms and standing to 3 meters, giant beavers weighing 200 kilograms and nearly ten times the size of modern beavers, mastodons and mammoths up to 4 m at shoulder height and weighing up to 10 tons. The museums of the world are crowded with the remains of some of these animals.

Why did these large animals disappear? There are two broad hypotheses of the causes of these extinctions. The anthropogenic or "overkill" hypothesis suggests that early humans caused the extinction of these large animals by a combination of hunting and the habitat changes brought about by burning the vegetation. The alternative hypothesis suggests that rapid climate change doomed large animals to extinction. The majority of the evidence now favors the human factor as the dominating cause of these extinctions (Zuo et al. 2013). Figure 7.5 shows schematically how these extinctions occurred at different times on different continents, typically coinciding with the arrival of humans during the late Pleistocene.

Australia's main extinction events occurred much earlier than those of the Northern Hemisphere. The picture in South America is less clear because the extinction events occurred over a longer period of time than they did in North America (Barnosky and Lindsey 2010).

The bottom line is that the large animals we see in the world today are a small subset of a much larger set of large mammals and birds that disappeared relatively recently due largely to our role as a major keystone species that could prey on any part of the food web.

WOLVES IN YELLOWSTONE

Wolves were virtually exterminated in the western United States by the 1950s, and pressure developed to reintroduce wolves to the large national parks from which they were lost. The Yellowstone ecosystem was a prime choice for reintroduction because of its large size, and in 1995–96 grey wolves from Canada were released in the park to help ameliorate ecological problems that had developed since the wolf was removed. After the last Yellowstone wolf was eliminated in 1926, park biologists became concerned about the impacts of elk browsing on vegetation in the elk winter range. The park managers undertook a program of elk culling that lasted from the mid-1930s until 1968. After a policy change by the park stopped the culling of elk, the elk population increased rapidly from an estimated low of just over 3,000 in 1968 to a high of about 19,000 in 1994. Elk have been the primary prey of the introduced wolves, and in addition grizzly bears commonly kill elk calves and scavenge on carcasses of elk and other ungulates killed by wolves. Figure 7.6 shows how elk have declined since the wolf reintroduction.

Although the wolves had a dramatic effect on elk density, the effects did not cascade down the food web of Yellowstone National Park as expected. The key effect of wolves in reducing elk numbers should allow the vegetation to recover. There has been concern about the impact of elk grazing on aspen trees. Aspen (*Populus tremuloides*) has been declining in large portions of the western US since the 1920s. The causes of this decline have become topics of much scientific investigation, and elk browsing has been one suggested cause. The decline of aspen is of conservation concern for many reasons, including the fact that aspen provides important habitat for elk as well as songbirds.

Food web dynamics are affected by many physical factors that must also be taken into account in explaining the failure of aspen recruitment. Fire, for example, can have strong effects on the recruitment of aspen seedlings

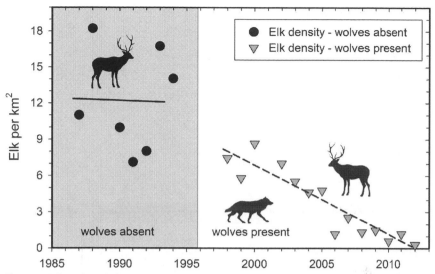

Figure 7.6 Density of elk in the eastern portion of the northern range of Yellowstone National Park before and after the reintroduction of the grey wolf. (Data from Ripple et al. 2014a.)

and in some ecosystems be more important that elk browsing (Eisenberg et al. 2013). Rainfall and temperature affect plant recruitment, and the simple idea that aspen recruitment depends only on elk browsing needs to be expanded to accommodate the effects of other environmental drivers. There is no question, however, that very dense ungulate populations, produced by human elimination of large predators like wolves and cougars, produce strong effects on plant production and thus on other species in the community (Teichman et al. 2013). The human factor operates as a strong driver of food webs now just as it did in the Pleistocene (Figure 7.5). Reintroducing wolves is a useful management tool for national parks but it comes at a cost to cattle ranchers outside the park, and the spatial scale of movements of large predators like wolves compounds problems of ecosystem management in areas with extensive human land use for agriculture.

LARGE CARNIVORE CONSERVATION

The largest terrestrial carnivores are prime candidates for keystone species. They are all wide-ranging and rare and occupy a position at the top of food webs. They are also some of the most imperiled species. Most large carnivores have experienced substantial population declines throughout the world during the past 200 years. Their food requirements and wide-

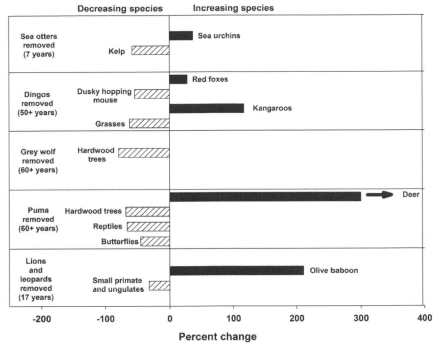

Figure 7.7 Five examples of community changes after the removal of large-carnivore species and the number of years since large-carnivore extirpation. The percentage change in abundance of each species group was calculated by dividing the value of each without predators by those with predators present. Black bars to the right indicate a positive effect of large carnivore removal, and striped bars to the left indicate a negative effect of removing large carnivores. Deer have responded so dramatically to puma (mountain lion) removal that they are off the scale to the right. References for these data are given in Ripple et al. (2014b). (Modified from Ripple et al. 2014b.)

ranging behavior often bring them into conflict with humans and livestock, as we saw above with the grey wolf in Yellowstone. This, in addition to human intolerance, renders them vulnerable to extinction.

Large carnivores face extinction threats because of habitat loss, persecution, and depletion of their prey. Because of their impacts on community structure, many of these large carnivores are candidates for keystone species status. Five of these species (Figure 7.7) show clear impacts characteristic of keystone species. The maintenance or recovery of ecologically effective densities of large carnivores is an important conservation objective for maintaining the structure and function of diverse ecosystems. Human management activities cannot easily replace the role of large carnivores.

The classic conception of large-carnivore influences on communities was that predators were responsible for depleting valuable resources such as fish and domestic livestock. This assumption is still used to justify wildlife management practices aiming to reduce or eradicate predators in some areas. This conception of carnivore ecology is now outdated. The roles large carnivores play in ecological communities are complex and varied, and their myriad social and economic effects on humans include many benefits that more than compensate for the losses of livestock. Conservation decisions must begin to account for the important ecological roles of large carnivores and the economic costs of carnivore species losses.

CONCLUSIONS

Determining the relative importance of species in food webs is difficult. Biological communities may have many redundant species that can be lost with little effect on community structure and function. Dominant species that make up the majority of the numbers or biomass in a community are often replaceable if they should disappear. This is contrary to our basic feeling that the larger the numbers, the more important a species must be. Keystone species are often relatively rare or of low biomass in a community; nevertheless, if they are removed, a drastic reorganization can be expected. Keystone species are best recognized experimentally by exclusion experiments, and may be more common in aquatic ecosystems than in terrestrial ones. Keystone species may be of major importance in restoration ecology if they can be reintroduced into areas from which they have been lost. Many large carnivores are top predators and have the potential for keystone effects on communities if they are removed by humans. The wolf is a prime example.

CHAPTER 8

NATURAL SYSTEMS ARE PRODUCTS OF EVOLUTION

KEY POINTS

- Evolution has occurred in the past and continues to occur, but now is strongly affected by human activities in agriculture, forestry, and medicine, and by climate change.
- The use of herbicides in agriculture, antibiotics in medicine, and poisons in pest control have all resulted in the evolution of resistant weeds, disease organisms, and pests like rats. Management of these problems must be changed to prevent further evolution of such resistant organisms in all these areas.
- Plants and animals can adapt to climatic warming by genetic changes, but for large species natural selection operates slowly while climate change is happening quickly. We know too little about the scope of genetic variation to predict how optimistic we can be about the ability of the Earth's biodiversity to cope with changing climates.

The sweep of evolutionary history should humble ecologists; humans have existed on Earth for only a very short time, and yet we have had a disproportionate impact on what has evolved over eons. Evolution has laid the groundwork for what we see today in the animal and plant life on different continents, and our concern here is how humans have altered and are altering the path of evolution in both positive and negative ways.

Can we use our improving knowledge of how modern communities evolved in the past to help us design better systems of forestry and agriculture? How much are organisms constrained by past evolutionary events? Are we encouraging pest species with our short-term fixes for controlling pests or managing disease outbreaks? These are all important questions at the interface between ecology and evolution, and an active area of research in biology.

Evolution is descent with modification, as Charles Darwin put it, or the cumulative change in characteristics of organisms over many generations. Darwin's genius was to recognize that the mechanism for evolution

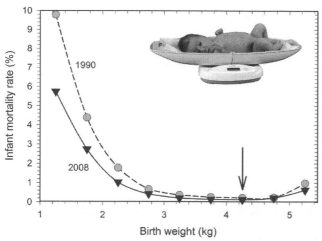

Figure 8.1 Natural selection for birth weight in humans. Data from United States for infants for 1990 and 2008. The optimal birth weight (arrow) is 4.25 kilograms, with a broad range of minimal mortality between 3.2 kilograms and 4.8 kilograms. Because of medical advances, infant mortality has been falling steadily, so the 1990 curve is higher than the 2008 curve. (Data from Mathews and MacDorman 2012.)

is natural selection. Natural selection is a process that arises from three basic observations: (1) variation among individuals in some attribute (such as eye color or body size), (2) consistent differences in reproduction or survival among individuals having this attribute, and (3) inheritance, or a resemblance between parents and offspring for the attribute. The process of natural selection operates to produce organisms that are adapted to the environments in which they live. Favorable attributes are those that give higher reproductive rates or better survival. Individuals with such favorable attributes are "selected for," and thus these attributes become more common in the population over several generations. But natural selection does not produce one "perfect" model for each species because the environments in which species live are not uniform but highly varied.

Humans are also subject to natural selection, although many might think we have progressed beyond this limitation. A good example is selection associated with birth weight in humans. Figure 8.1 shows that early infant mortality in the United States is lowest for babies weighing about 4.2 kilograms, slightly above the observed mean birth weight of 3.4 kilograms for the population. Very small babies die more frequently and therefore cannot pass on their genes, and very large babies are at increased risk even with modern medical care. Natal death rates have been falling in the US but have changed only a little since 2000 in spite of advances in medical care in the last 14 years.

EVOLUTION OF RESISTANCE TO POISONS IN RODENTS

Evolution can work to our disadvantage if we are not careful, and this is the history of rodent control with anticoagulants. The main poisons used to control rats in cities and on farms are anticoagulants. The first-generation anticoagulant was warfarin (patented in 1948 by the Wisconsin Alumni Research Foundation, or WARF). Warfarin prevents mammalian blood from coagulating and in minute doses it was discovered to have a great medical benefit by preventing blood clots in people susceptible to strokes. In larger doses it is toxic to mammals and birds. Warfarin in its various trade names was considered to be the solution to the problem of rats in cities and farms, but already in 1958 Norway rats (*Rattus norvegicus*) were found in Wales that were resistant to warfarin (Buckle 2013). Five mutations have been discovered that confer resistance to anticoagulants in Norway rats in the United Kingdom, but in continental Europe only one or two genetic mutations seem to be involved. The development of resistance to first-generation anticoagulants has occurred independently in different localities in Europe (Figure 8.2).

The management response to warfarin resistance in rats in the 1950s and 1960s was the development in the 1970s of second-generation rodenticides that are much more powerful, and thus more dangerous to use. The important issue now is whether or not Norway rats and house mice are developing resistance to second-generation rodenticides as well (Table 8.1) (Pelz et al. 2012). In England, for example, Norway rats, black rats, and house mice are now resistant to the second-generation rodenticides, making it difficult to control rats in cities and farms (Buckle 2013).

Anticoagulant rodenticides can still be effective for the control of Norway rats over large parts of the Europe. However, resistance to anticoagulants is widespread and is an increasing threat to effective rat control. Regulatory authorities in different countries are concerned about the potential environmental impacts of anticoagulant rodenticides. The occurrence of residues of second-generation anticoagulants in wildlife has been increasing. For example, about 90% of barn owls (*Tyto alba*) in England carry residues of rodenticides, and residues in kestrels (*Falcon tinninuculus*) and red kites (*Milvus milvus*) are almost as frequent there (Buckle 2013).

The use of poisons in controlling pests has been likened to an arms race

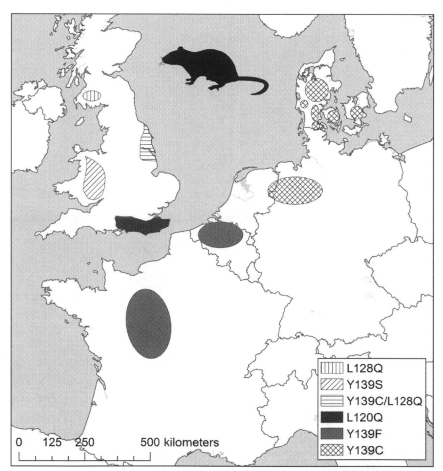

Figure 8.2 Geographic origin of rodent populations resistant to anticoagulants in Western Europe. Warfarin resistance areas where resistant rats were trapped in the wild are indicated. Different mutations conferring resistance are represented by different patterns. See also Table 8.1. (From Pelz et al. 2005.)

in which we develop new poisons for control and the animals or plants develop resistance over time, leading to yet another round of developing different poisons, followed by yet more resistance. It is a good example of evolution in action over a short time period.

HERBICIDE RESISTANCE IN WEEDS

Farming systems around the world have changed dramatically during the last 70 years, partly because of rising human populations but more

Table 8.1 Resistance to second-generation anticoagulants in Norway rats (*Rattus norvegicus*), black rats (*Rattus rattus*), and house mice (*Mus musculus*) in different countries in 2005. (From Pelz et al. 2005).

Country	Norway rat	Black rat	House mouse
Belgium	+		+
Denmark	+	+	+
Finland			+
France	+	+	+
Germany	+	+	+
United Kingdom	+	+	+
Italy	+		
Netherlands	+		+
Sweden			+
Switzerland			+
Canada	+		+
United States	+	+	+
Japan		+	
Australia		+	

because of innovations in agricultural science. Weed management was one important innovation after 1945, when general herbicides were developed. The simple option was to spray the weeds in a field before planting the crop. But a more attractive option became available once crops could be genetically selected that were resistant to herbicides. Many of the major crops were developed in the 1990s to be herbicide resistant, and the general approach would then be to sow the resistant crop and then spray it later with herbicide, thereby killing all the weeds but not affecting the crop. For a short time, such herbicides were very effective and this strategy worked, but now it is not working very well because of continuing natural selection for herbicide-resistant weeds.

One of the major problems of every agricultural cropping system is weeds. A number of studies have suggested weeds have the greatest potential for yield loss with estimates of actual losses of about 10% worldwide. The percentage crop yield loss attributable to weeds has changed little since the 1960s, suggesting that crop protection companies, crop breeders, farmers and weed biologists are locked in a weed management arms race. The evolution of resistance to herbicides in weeds provides another example of evolution in action. Since the 1960s, when herbicide resistance was first reported, resistance to a broad range of herbicides has been confirmed in 189 weed species (Neve et al. 2009). In some weeds, populations have evolved multiple resistance, in which resistance to one herbicide mode of action has necessitated a switch to other kinds of herbicides to which resistance has subsequently evolved through multiple independent mechanisms. The arms race is being lost because the rate of discovery of new herbicides has declined while the evolution of herbicide resistance has increased. As weed control technologies advance, selection for more resistant weeds will intensify.

Glyphosate was developed in the 1970s, and once glyphosate-tolerant crops were developed in the 1990s, glyphosate was deemed to be the beginning of a new era in weed control. Glyphosate (N-(phosphonomethyl)-glycine) was promoted as a perfect herbicide, better than any other herbicide currently available, and is the world's best-selling herbicide. Glyphosate kills nearly all herbaceous plants. By contrast, it has been considered harmless for animals, including humans. Glyphosate has been assumed to be rapidly inactivated in soils by microbial degradation. Because it targets a metabolic pathway that occurs only in plants, glyphosate has been proclaimed to be safe to nontarget organisms. It enables no-till cropping that decreases erosion and nutrient leaching without disturbing soil structure and functions. It can also be used as a defoliant in forestry and to control invasive species in the context of conservation weed management. Those parts of the world that have adopted genetically modified glyphosate-resistant crops are now experiencing high levels of evolved weed resistance to glyphosate (Helander et al. 2012). And all of the safety claims for glyphosate may be overstated.

To investigate the rate of change in herbicide-resistant weeds, two major surveys of cereal crops in 7.7 million hectares of cropland in the Canadian prairie region were carried out from 2001 to 2011 (Beckie et al. 2013). Figure 8.3 illustrates the rapid rate of increase in resistance to the main herbicides in use in this large farming area. The propensity for evolution of resistance varies, with some weed species being more prone to resistance than others (Heap and LeBaron 2001). In the most extreme cases, resistance has evolved following exposure of no more than three or four generations of a weed population to a particular herbicide. Herbicide resistance is arguably the single largest global weed management issue.

Farmers are able to counter herbicide resistance by using a variety of

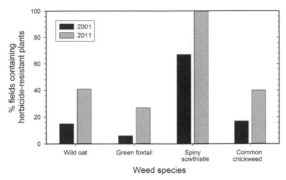

Figure 8.3 The change from 2001–2011 in the percentage of agricultural fields in the Canadian prairie provinces that contained herbicide-resistant weeds. In this ten year period herbicide resistance in weeds nearly doubled. The species are wild oat (*Avena fatua*), green foxtail (*Setaria viridis*), spiny sowthistle (*Souchus asper*), and common chickweed (*Stellaria media*). (Data from Beckie et al. 2013.)

measures of weed control, and by now it is clear that reliance on weed control only with herbicides will not be sufficient, even if a different mix of herbicides is used each year. Physical controls (such as the mowing of weeds) or improved agricultural practices (such as avoiding monoculture crops) can help to control weeds. A good example of the variety of methods that can be applied to weed management is the set now developed for rice crops in Southeast Asia (Chauhan 2012).

Rice is a principal source of food for more than half of the world's population, and more than 90% of rice worldwide is grown and consumed in Asia. Dry-seeded rice is replacing manual transplantation of seedlings as the established method of planting rice in some countries as growers respond to increased costs or decreased availability of labor or water. However, weeds are a major constraint to rice production because of the absence of the size difference between the crop and the weeds at crop establishment. Herbicides are used to control weeds in dry-seeded rice, but because of concerns about the evolution of herbicide resistance and a scarcity of new herbicides, there is a need to integrate other weed management strategies with herbicide use. In addition, different weed species have variable growth patterns so that any single method of weed control cannot provide effective control in rice crops. The adopted strategies include tillage systems in which the rice field is irrigated before the rice is sown so the weeds germinate and then are killed by cultivation; the use of weed-competitive rice varieties; controlled water depth after crop emergence; optimum row spacing; manual weeding; and appropriate herbicide timing. Herbicides are an important component of weed control in rice cultivation but they are not the only methods (Chauhan 2013).

ANTIBIOTIC RESISTANCE IN MICROORGANISMS

The introduction of antibiotics in the last century was a critical factor in the reduction of human mortality. This benefit has been compromised by the continuing overuse of antibiotics in medicine and in food production, which has selected for resistance among human pathogens. High frequencies of resistance significantly reduce the possibility of effectively treating infections, which increases the risk of complications and fatal outcomes. The development of antibiotic resistance increases the economic burden on the health care system and must be reversed if we are to maintain the utility of antibiotics.

Since the introduction in 1937 of sulfonamides (or sulfa drugs), the first effective antimicrobials, the development of specific mechanisms of resistance has plagued their therapeutic use. Sulfonamide resistance was originally reported in the late 1930s, and the same mechanisms operate 80 years later. Penicillin was discovered by Alexander Fleming in 1928, and in 1940, several years before the introduction of penicillin as a therapeutic drug, bacteria were discovered that were preadapted to be resistant to penicillin. Once the antibiotic was used widely in the 1940s, resistant strains became prevalent. Streptomycin was introduced in 1944 for the treatment of tuberculosis and almost immediately mutant strains of the causative agent *Mycobacterium tuberculosis* resistant to therapeutic concentrations of the antibiotic were discovered. As other antibiotics have been discovered and introduced into clinical practice, a similar course of events has ensued. Figure 8.4 shows chronologically the sequence of discovery and resistance development for antibiotics.

The unexpected identification of genetically transferable antibiotic resistance in Japan in the mid-1950s changed the whole picture by introducing the heretical genetic concept that collections of antibiotic resistant genes could be disseminated by bacterial conjugation throughout an entire population of bacterial pathogens. Only in the past few years has it been appreciated that gene exchange is a universal property of bacteria that has occurred throughout eons of microbial evolution. The discovery of the presence of bacterial gene sequences in birds and mammals has heightened awareness of the great importance of gene transfer between

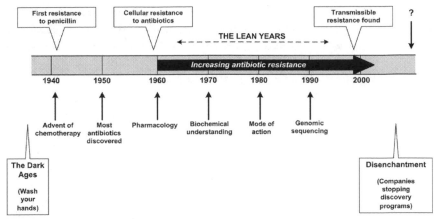

Figure 8.4 History of antibiotic discovery and the development of antibiotic resistance. In the dark ages before 1940, when the *only* way to avoid infection was wash your hands. 1950 marked the golden era, when most of the antibiotics used today were discovered. This was followed by the lean years, the low point of new antibiotic discovery and development. By 1960, pharmacology attempted to improve the use of antibiotics by adjusting doses. By 1970, biochemical knowledge of the actions of antibiotics and the resistance mechanisms led to chemical modification studies to avoid development of resistance. By 1980, mode-of-action and genetic studies led to efforts to design new antibiotics. In 1990, genome sequencing methods were used to predict essential targets for antibiotics. This was followed by an era of disenchantment, the failure of the great investment in genome-based methods that led many companies to discontinue their discovery programs. Before antibiotics were discovered, Ignaz Semmelweis in 1847 advocated hand washing as a way of avoiding infection; this practice is now strongly recommended as a method to prevent transmission. (From Davies and Davies 2010.)

unrelated species in evolution. Gene transfer occurs between different bacteria species, and the same genes for pathogenicity have been found in different bacterial genera.

Many of the bacterial pathogens associated with epidemics of human disease have evolved into multidrug-resistant forms subsequent to antibiotic use. Multidrug-resistant forms of the bacteria that causes tuberculosis (*M. tuberculosis*) are major pathogens found in both developing and industrialized nations; they have become the 21st-century version of an old deadly pathogen. Other serious infections include hospital-linked infections with an array of bacteria that are now referred to as "superbugs," bacteria that cause enhanced likelihood of mortality because of multiple mutations that endow them with high levels of resistance to all the available antibiotics.

The appearance and dissemination of antibiotic-resistant pathogens have stimulated countless studies of the genetic aspects of resistance development. Acquisition of resistance has long been assumed to incur a serious energy cost to the microorganism so that resistant strains would lose out in competition with nonpathogenic strains of the same species. Resistant pathogens typically have lower growth rates than nonresistant clones of the same bacteria. As a result of this reduction in fitness, multidrug-resistant strains were suggested to be unstable and short-lived in the absence of selection so that they could not spread to healthy people (Andersson 2006). If these findings are correct, they imply that resistance could be reversible, provided antibiotic use is reduced. In this way the problem of superbugs would solve itself. Unfortunately, this has not turned out to be correct.

Recent studies have shown that resistant microbes are unlikely to disappear even if we reduce antibiotic use (Davies and Davies 2010). Resistant bacteria are able to persist for a long time, even if no antibiotic selective pressure is present. These findings emphasize the need for the development of new antibiotics and the importance of prudent antibiotic use to reduce resistance development. The evolution of resistance to drugs is a textbook example of rapid evolution, and is a major issue for public health. The number of drugs with different mechanisms of toxicity acting on independent bacterial targets has proved to be limited. Only a few new active molecules have been discovered during the past 30 years, underlining the need for careful antibiotic use in medicine (Bourguet et al. 2013).

MYXOMATOSIS IN THE EUROPEAN RABBIT

The European rabbit (*Oryctolagus cuniculus*) was introduced into Australia in 1859 and increased to very high densities within 20 years. After World War II, an attempt was made to reduce rabbit numbers by releasing a viral disease called myxomatosis. Myxomatosis originated in the South American jungle rabbit (*Sylvilagus brasiliensis*). In its original host, myxomatosis is a mild disease that rarely kills its host, but in the European rabbit, a new host, it was lethal. Transmission of myxomatosis occurs via biting arthropod vectors, principally mosquitoes and fleas.

When it was first introduced into Australia in 1950, myxomatosis was highly lethal to European rabbits, killing over 99% of individuals infected (Figure 3.4, page 35). Myxomatosis was also introduced to France in 1952, from which it spread throughout Western Europe, reaching Britain in 1953. In Britain 99% of the entire nation's rabbit population was killed in the first epizootics from 1953 to 1955.

Very soon after its introduction, weaker myxoma virus strains were detected in England and in Australia (Fenner and Ratcliffe 1965). Since myxomatosis was introduced into Britain and Australia, evolution has been going on in both the virus and the rabbit. The virus has become weaker, such that it kills fewer and fewer rabbits and takes longer to cause death. Because mosquitoes are a major vector of the disease, the time period between exposure and death is critical to viral spread, and weaker viruses are at an advantage for spreading. By testing standard laboratory rabbits (domesticated European rabbits) against the virus, virologists can measure viral changes while holding rabbit susceptibility constant. Since 1951, less virulent grades of virus have replaced more virulent grades in field populations. At the same time, rabbits have also become more resistant to the virus. By challenging wild rabbits with a constant laboratory virus source, virologists can detect that natural selection has produced a growing resistance of rabbits to this introduced disease.

There has been considerable speculation that all diseases will behave like myxomatosis and show reduced virulence as time passes. This comforting idea has now been rejected (Ebert and Bull 2003). Many scientists have raised concerns about the threat posed by human intervention on the evolution of parasites and disease agents (Mennerat et al. 2010). New parasites (including pathogens) keep emerging, and parasites that previously were considered under control are reemerging, sometimes in highly virulent forms. This reemergence may be parasite evolution driven by human activity, including ecological changes related to modern agricultural practices. Intensive farming creates conditions for parasite growth and transmission drastically different from what parasites experience in wild host populations and may therefore alter natural selection on virulence. Although recent epidemic outbreaks highlight the risks associated with intensive farming practices, most discussion has focused on short-term economic losses imposed by parasites such as costly applications of toxins for chemotherapy of livestock. Ecologists have predicted the evolutionary consequences of intensive farming by relating current knowledge of parasite life-history and virulence with specific conditions experienced by parasites on farms. Intensive farming practices are likely to select for fast-growing, early-transmitted, and hence more virulent parasites.

The global fish farming industry, which has expanded 10-fold during the last 25 years, illustrates this problem all too well. For migrating marine fish species like salmon, this enormous increase in large populations concentrated in pens is associated with another change. Instead of a seasonal presence of migratory fish like salmon in summer, there is now a year-round presence of salmon in pens in coastal seawater, which provides a highly predictable resource for parasites and a crowded population for easy transmission. Increasingly virulent strains of bacterial parasites of salmon and trout in Finland have followed from attempts to treat the farmed fish with antibiotics (Mennerat et al. 2010).

The response of the fish farming industry has been to treat the fish farms with antibiotics, insecticides, or other toxins that reduce the parasite population. Sea lice (*Lepeophtheirus salmonis*) on farmed salmon have been a particularly controversial issue (Peacock et al. 2012). Figure 8.5 illustrates how the parasitic load on wild pink salmon in an area with extensive fish farming has fluctuated over 9 years. Throughout this period the fish farms treated their caged salmon with a parasiticide (emamectin benzoate), and it was effective in reducing sea lice abundance both in farm fish and in the wild salmon migrating by the fish farms. The difficulty is that sea lice have rapidly evolved resistance to widely used antibiotics and insecticides used for treating farmed fish. We appear to be repeating in fish farms the same problems we have generated with microbial resistance to antibiotics used to treat human diseases. Sea lice are now a major problem wherever salmon are farmed, and the short-term solution of using toxins to control sea lice is leading the industry into another arms race with the parasites and pathogens of farmed fish.

EVOLUTIONARY ADAPTATION TO CLIMATE CHANGE

Organisms can adapt to any environmental change either by tolerating changed conditions or by changing their genetic makeup by natural selection favoring the most fit individuals. For example, as the ocean grows more acidic because of rising carbon dioxide levels in the air, organisms can simply tolerate the increased acidity or, by selection, only those that can tolerate increased acidity will survive, changing the genetic composition of the population. Genetic change is possible only if there are genetic variants for the ability to tolerate the changed conditions (Bell 2013). If this is the case, we can readily have evolutionary rescue and we will see no change in life as we know it—all the changes will be in the genome of the adapted individuals. This is the optimist's view of climate change—that little will happen and alarm bells need not go off with the changes we are now experiencing. It is possible to build mathematical models that show such genetic changes are possible.

The pessimist's view relies on two historical observations. First, al-

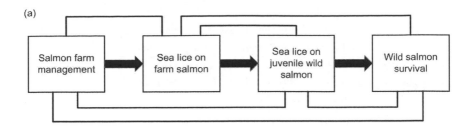

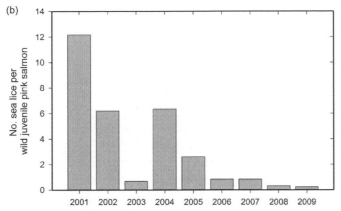

Figure 8.5 (a) The origin of the sea lice problem in salmon farming areas of coastal British Columbia. Arrows indicate scientific studies that have established this causal chain. (b) Mean abundance of sea lice on wild juvenile pink salmon in the Broughton Archipelago of British Columbia in 2001–2009. During this time salmon farms were treating their penned fish with an insecticide to kill sea lice. The number of treatments increased rapidly after 2004, and the infestation rate on wild salmon caught in the open ocean near the pens also declined. In 2003 the government shut down the salmon farms for one year, greatly reducing sea lice in this area and reducing the infestation rate in wild salmon. The problem is solved until sea lice become resistant to the pesticides being used. (From Peacock et al. 2012.)

though organisms in the past have adapted to climate changes, this adaptation has occurred over thousands to tens-of-thousands of years at a very slow rate. The current climate change scenario is extremely rapid by comparison, operating on a time scale of at most hundreds of years, perhaps on average about 100 times as fast as what has happened in geological time. While microbes with fast generation times might adapt to current changes, most of the biodiversity we see on Earth will not be able to adapt in a short time and will potentially disappear.

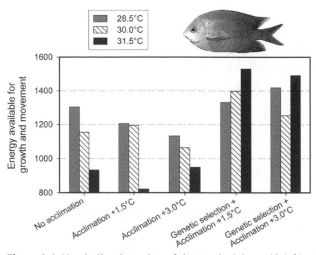

Figure 8.6 Metabolic adaptation of the tropical damselfish (*Acanthochromis polyacanthus*) to rising ocean temperatures. Fish were raised at normal sea water temperature (28.5°C) and at higher temperatures (30.0°C and 31.5°C) for 2 generations to test for the ability to adjust their metabolism at sea water temperatures that are predicted to occur in 2050 to 2100 AD. Fish were also selected for 2 generations for performance at the higher temperatures. The results show that genetic selection could adjust temperature tolerances to preserve metabolic energy levels that would permit survival in the warmer ocean water. (Modified from Donelson et al. 2012.)

The second less comforting observation is that most organisms that have been studied have defined limits of tolerance to environmental change: to temperature, rainfall, and nutrients for terrestrial organisms and to temperature, acidity, and nutrient levels for aquatic organisms. This fact is obvious to all people who farm or garden—flowers do not bloom in cold winters and wheat does not grow if there is no soil moisture. The optimist replies with data like that shown in Figure 8.6. A coral reef fish that is adapted to sea water temperatures of 28.5°C was tested at two higher temperatures in two ways (Donelson et al. 2012). Adaptation can occur by having a wide temperature tolerance, or by genetic selection of parents that do better in warmer water. Both these forms of adaptation operate in this coral reef fish species, showing that it can tolerate rising ocean temperatures expected during the next 100 years. But if ocean temperatures continue to rise in the coming centuries, the limits of tolerance could be exceeded and adaptation will fail.

These kinds of studies are important, but they answer only part of the overall question first because they must be carried out on each species in the ecosystem separately, an impossible task for rich tropical ecosystems, and secondly because they must be done in the laboratory without all the other species interactions that occur in wild populations—interactions with food supplies, predators, parasites, and diseases. Climate stress is only one stress to which species must respond to survive.

Some species can evolve to cope with increasing climatic stress over rapid time scales and thus evolution can rescue these populations from extinction. The species for which rescue through evolution will be most probable are species with very large populations and short generation times (microorganisms), but these groups of species are at the lowest risk of extinction. The critical species with the greatest risk of extinction from climate change are large species with long generation times and small population sizes. These large species are the most difficult to study if only because of the time required for one generation, and for these we currently lack critical information about the genetic variation that is available to respond to climate change.

The rate of environmental change will affect the capacity of all populations to persist long enough to adapt to climate change. This means at the operational level that we must take global action now to slow the rate of climate change because this will diminish its impacts and maximize the potential for some species to adapt.

CONCLUSIONS

The fact that evolution has occurred in the past and will continue to occur in the future carries with it a two-sided message. First, humans can interfere with evolution through the use of chemicals that are beneficial in medical science, agriculture, and forestry. But when we do this we need to follow the Hippocratic Oath for the environment, "Scientists, do no harm." The negative consequences that have followed from excessive pesticide use, antibiotic misuse, and herbicides in agriculture are just three important problems of our day that have their roots in evolution.

The most critical problem we now face is that of climate change and the possibility of a great extinction event that could lead to an Earth that is uninhabitable by humans. We and the other species on the Earth might be able to adapt and evolve tolerance to the slow changes that are now occurring, but this is far from a certainty. We should not assume that rapid rates of evolution will solve the climate change issues of our day. There is a certain irony in that rapid evolution has in fact produced many of the problems we have highlighted in this chapter. There are many important actions that should follow from these simple evaluations, and while the science is relatively clear, the social and political actions that should follow are not yet in sight. Conducting uncontrolled experiments with the Earth's ecosystems is not to be recommended.

CHAPTER 9

NATURAL SYSTEMS RECYCLE
ESSENTIAL MATERIALS

KEY POINTS

- Every ecological system is driven by nutrients in the long term, and in the short term we must conserve essential elements.
- The law of recycling is simple: input must equal output or the system decays. The environment is not a bank of infinite resources.
- Humans now have large impacts on the sulfur cycle (acid rain), the nitrogen cycle, the carbon cycle, and the water cycle. Reducing these impacts is essential for sustainable living.

All species depend on renewable resources for continued survival; consequently, the most essential physical process for biological communities is recycling. Through the long history of evolution, plant and animal communities have learned the hard lesson that every schoolchild learns after opening his or her first bank account: income must equal or exceed expenditure in the long run. Expenditure can exceed income for short times, but this imbalance can never occur over a long period without bankruptcy. Recycling theory is just this very simple point: for all the materials needed for life in an ecosystem input must equal output, or the ecosystem degrades in the long term. How do natural communities recycle essential materials to achieve balance, and what are the difficulties they face? The recycling of essential resources is important for humans because it is necessary to achieve sustainable living.

Plants and animals require two basic physical provisions: energy and materials. Energy in ecological systems originates from the sun, and all of the energy needed must be captured by green plants through photosynthesis. The sun's energy is captured in plant carbohydrates, proteins, and lipids, and then used by animals through the food web. Energy is not recycled by organisms, but is ultimately used for activity or growth and lost as heat. The continued input of solar energy keeps the ecological machine running, and organisms can afford to use energy because there is always more on the way.

Materials are completely different—they are not provided anew each

(a)

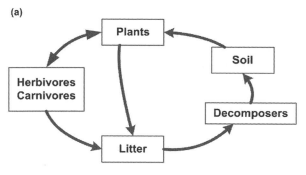

Local cycles of P, K, Ca, Mg, Cu, Zn, B, Cl, Mo, Mn, and Fe

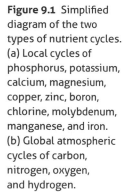

Figure 9.1 Simplified diagram of the two types of nutrient cycles. (a) Local cycles of phosphorus, potassium, calcium, magnesium, copper, zinc, boron, chlorine, molybdenum, manganese, and iron. (b) Global atmospheric cycles of carbon, nitrogen, oxygen, and hydrogen.

(b)

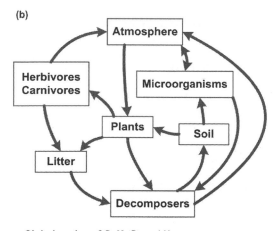

Global cycles of C, N, O, and H

day and must be conserved. Conservation can be achieved only by recycling, and so we need to find out how chemical materials cycle in nature. Two quite different cycles occur (Figure 9.1). Elements exchanged in a gaseous state—carbon, nitrogen, oxygen, and hydrogen—all circulate in global cycles because they are exchanged between the air and biological organisms. Long-distance transfers are common, and the oxygen you are breathing today is part of a pool or reservoir having many distant sources. Solid elements that are essential for living organisms—phosphorus, potassium, calcium, magnesium, copper, zinc, boron, chlorine, molybdenum, manganese, and iron—all circulate in local cycles (Figure 9.1) and have no mechanism for long-distance transfer.

To analyze any particular nutrient cycle, we need to measure first of all the amount of the nutrient in each "box" of Figure 9.1, and second the flow rate of the nutrient between boxes. Once we have completed this description, we can then try to determine the main factors affecting the transfer of nutrients in the community. Nutrient cycles have been studied in many biological systems, particularly in systems that humans exploit.

The nutrient dynamics of a forest involve an input of nutrients from soil and rock weathering, and an output of nutrients in leaf litter and root decomposition. Nutrients are also lost from a forest in the harvesting of logs. Forests, like all biological communities, are not closed systems. Animals move from one community to another, and drainage water can transport dissolved materials to adjacent lakes and streams. If commercial forestry is operating in a particular forest community, the logs removed from the forest represent a significant investment in nutrients. These losses must be counterbalanced by nutrient gains from rain and dust, soil and rock weathering if there is to be no long-term decline in forest productivity. Atmospheric inputs of nutrients are critically important in many ecosystems affected by human-caused air pollution as well as natural dust storms. Atmospheric inputs of phosphorus in desert dust is essential to maintaining productive tropical rainforests in the Amazon Basin (Bristow et al. 2010). These gains and losses have been studied in detail in only a few ecosystems.

NUTRIENT CYCLES IN THE HUBBARD BROOK ECOSYSTEM

One of the most extensive studies of nutrient cycling in forests has been carried out at the Hubbard Brook Experimental Forest in New Hampshire. The Hubbard Brook forest is a nearly mature hardwood forest ecosystem. The area is underlain by rocks that are relatively impermeable to water, and hence all runoff occurs in small streams. The area is subdivided into several small watersheds that are distinct yet support similar forest communities, and these watersheds are good experimental units for study and manipulation.

Nutrients enter the Hubbard Brook forest ecosystem in precipitation, and the precipitation input was measured in rain gauges scattered over the study area. Nutrients leave the ecosystem primarily in stream runoff, and this loss was estimated by measuring streamflows. For most dissolved nutrients, the streamwater leaving the system contains more nutrients than the rainwater entering the system. About 60 percent of the water that enters as precipitation leaves as streamflow, and most of the remaining 40 percent is transpired by plants or evaporated.

Annual nutrient budgets can be calculated for watersheds in the Hubbard Brook system, based on the difference between precipitation input and stream outflow. If we assume that nutrient budgets should be in equilibrium in this undisturbed ecosystem, the net losses must be made up by chemical decomposition of the bedrock and soil.

After obtaining background information for the intact watersheds, Gene Likens at Cornell and F. H. Bormann at Yale studied the effects of logging on the nutrient budget of a small watershed at Hubbard Brook. One 15.6 hectare watershed was logged in 1966, and the logs and branches were left on the ground so that nothing was removed from the area. Great care was taken to prevent disturbance of the soil surface to minimize erosion. For the first 3 years after logging the area was treated with an herbicide to prevent any regrowth of vegetation. This deforested watershed was then compared with an adjacent intact watershed.

Runoff in the small streams increased immediately after the logging; annual runoff in the deforested watershed was 41%, 28%, and 26% above the control in the 3 years after treatment. Detritus and debris in the stream outflow increased greatly after deforestation, particularly 2 to 3 years after logging. Correlated with this was a large increase in stream-water concentrations of all major ions in the deforested watershed. Nitrate concentrations in particular increased 40- to 60-fold over the control values (Figure 9.2). For two years the nitrate concentration in the stream-water of the deforested site exceeded the maximum safe level recommended for drinking water. Average stream-water concentrations increased 417% for calcium, 408% for magnesium, 1558% for potassium, and 177% for sodium in the two years after deforestation.

Nutrient losses after clear-cutting in forests can be recovered if the site is not disturbed by severe soil erosion. Rock weathering restores many of the important elements like calcium and potassium. Nitrogen is captured from the air by certain bacteria and algae and converted to nitrate in the soil. At least 60–80 years are required for nutrient recovery at Hubbard Brook, and consequently the present forestry practice of a 110- to 120-year rotation for cutting should allow the forest to recover completely between cuttings.

By looking at commercial forestry from a viewpoint of nutrient cycling we can help to recommend some management procedures in forestry. For example, bark is relatively rich in nutrients and hence lumbering operations ought to be designed to strip the bark from the trees at the field site and not at some distant processing plant. Similarly, the debris of branches and leaves left on the logged area, although unsightly, represent part of

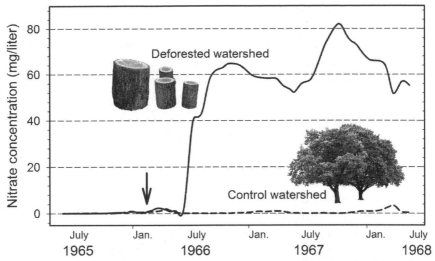

Figure 9.2 Stream water concentrations of nitrate in two watersheds at the Hubbard Brook Experimental Forest, New Hampshire. The arrow marks the completion of the cutting of the trees on the deforested watershed. The control watershed was not disturbed. Nitrate loss was greatly accelerated after logging. (The figure shows an increase of nitrate in the streams; this corresponds to the terrestrial nitrate loss.) Once vegetation grew back after several years, nitrate loss returned to control levels. (Modified after Likens et al. 1970.)

the nutrient capital of the site that will be recycled as it decomposes. Ecologically sound harvesting methods are not always aesthetically pleasing. Commercial forestry has many impacts on other species that use forests as habitats, so that many aspects of biodiversity conservation must be a priority of forest management, as well as nutrient cycling and nutrient retention.

Most of the nutrients lost from clear-cut areas end up in streams and lakes, where they become part of an aquatic pollution problem. Undesirable losses from some biological communities can become undesirable gains for another community. These flows of nutrients show very graphically how interdependent communities can be.

The main questions for which we do not yet have answers for forested ecosystems are whether nutrient cycling is being disrupted by forestry practices and whether air pollution from burning fossil fuels is affecting forests. The key element is nitrogen, an essential element for plant growth, and we need to look at the dynamics of nitrogen on a global scale.

GLOBAL NITROGEN CYCLE

The availability of nitrogen is often limiting for both plants and animals, both on land and in the oceans. Nitrogen is abundant in air (78% nitrogen), but few organisms can use this N_2 directly. A small number of bacteria and algae can use nitrogen from the air and fix it as nitrate or ammonia. Many of these organisms work symbiotically in the root nodules of legumes to fix nitrogen, and this is a major source of natural nitrogen fixation. Human additions to the global nitrogen cycle have become substantial, particularly with the production of nitrogen fertilizers for agriculture.

Nitrogen is a key element in modern agriculture, and fertilization of agricultural crops with nitrogen compounds greatly increases yield. Figure 9.3 illustrates how corn production in Iowa soils can be increased with added nitrogen in fertilizer. The key point is that there is an upper limit to plant growth at which point additional nitrogen has no effect. In this example (Figure 9.3), adding more than 150 kilograms of nitrogen per hectare generates no increase in crop production, and the excess nitrogen is lost to the soil, running off into rivers and lakes.

The global nitrogen cycle is strongly affected by humans. Human activities add about the same amount of nitrogen to the biosphere each year as do natural processes, but this human addition is not spread evenly over the globe. The impact of human additions of nitrogen has shown up particularly in changes in the composition of the atmosphere. Nitrogen-based trace gases—nitrous oxide, nitric oxide, and ammonia—have major ecosystem impacts. Nitrous oxide is unreactive chemically and long-lived in

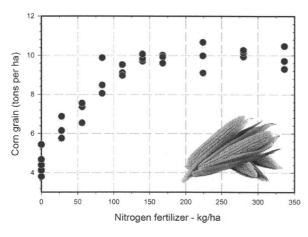

Figure 9.3 Increased corn production as a result of nitrogen fertilizer application in Iowa corn (maize) fields. No improvement is achieved if more than 150 kilograms per hectare is applied. (Data from Cerrato and Blackmer 1990.)

the atmosphere. It traps heat and thus acts as a greenhouse gas to change climate. It is increasing in the atmosphere at 0.25% per year. Nitric oxide, by contrast, is highly reactive and contributes significantly to acid rain as well as smog. Nitric oxide can be converted to nitric acid in the atmosphere, and in the western United States acid rain is based more on nitric acid than on sulfuric acid.

In the presence of sunlight, nitric oxide and oxygen react with hydrocarbons from auto exhaust to form ozone, the most dangerous component of smog in cities and industrial areas. Nitric oxide is produced by burning fossil fuels and wood. The third nitrogen-based trace gas in the atmosphere is ammonia. Ammonia neutralizes acids and thus acts to reduce acid rain. Most ammonia is released from organic fertilizers and domestic animal wastes. Domestic feedlots for the fattening of cattle are a major source of ammonia.

The effect of human activities on the nitrogen cycle has been an increased deposition of nitrogen on land and in the oceans. Since nitrogen additions are typically coupled with phosphorus additions, the result is increased algal growth in freshwater lakes and rivers and coastal marine areas. Phosphorus additions to freshwater typically increase primary production—the production generated by plants—while nitrogen addition to estuaries increases primary production in marine environments. The North Atlantic Ocean Basin receives nitrogen from many rivers that transport excess nitrogen into the ocean. Nitrogen input to the North Atlantic has increased 2- to 20-fold since 1750, and inputs from northern Europe are the highest. Nitrates in rivers are increasing throughout the Northern Hemisphere in proportion to the human population along the rivers. In the Mississippi River, nitrates have more than doubled since 1965. Groundwater concentrations of nitrate are also increasing in agricultural areas, and in some areas are approaching the maximum safe level of nitrate in drinking water (10 milligrams per liter). The results of nitrogen additions to aquatic systems are nearly all negative, reducing water quality.

Terrestrial deposition of nitrogen is unevenly distributed in industrialized countries. For example, in the USA nitrogen deposition is very low in the Western states and very high in the upper Midwest and Eastern states (see National Atmospheric Deposition Program: http://nadp.isws.illinois.edu). Nitrogen oxides originate from the burning of coal and oil products and are deposited across the landscape in rain and snow. The addition of nitrogen to terrestrial ecosystems can have positive effects. Nitrogen deposition on land can relieve the nitrogen limitation of primary production

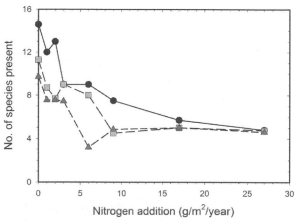

Figure 9.4 Vegetation responses to 12 years of nitrogen fertilization in Minnesota grasslands. Three fields were used, and 6 different plots were measured for each level of nitrogen addition. The number of species declines dramatically from 10–15 to 5 as more nitrogen is added to these grasslands. A few dominant species take over the grassland when nitrogen is not limiting. (Data from Wedin and Tilman 1996.)

that is common in many terrestrial ecosystems. Swedish forests are all nitrogen limited and have averaged 30% greater growth rates in the 1990s than the 1950s. The important concept here is that of the critical load — the amount of nitrogen that can be input and absorbed by the plants without damaging ecosystem integrity. When the vegetation can no longer respond to further additions of nitrogen (Figure 9.3), the ecosystem reaches a state of nitrogen saturation, and all new nitrogen moves into groundwater or streamflow or back into the atmosphere. Nitrate is highly water soluble in soils, and excess nitrate carries away with it positively charged ions of calcium, magnesium, and potassium. Excess nitrate can thus result in calcium, magnesium, or potassium limiting plant growth, and this is why most commercial garden fertilizers contain more than just nitrogen.

Increasing nitrogen in terrestrial ecosystems can also have undesirable impacts on biodiversity. In most cases, adding nitrogen to a plant community reduces the biodiversity of the community. Figure 9.4 illustrates the impact of experimentally adding nitrogen for 12 years to grasslands in Minnesota. Species that are nitrogen responsive, often grasses, can take over plant communities enriched in nitrogen. The Netherlands has the highest rates of nitrogen deposition in the world, largely due to intensive livestock operations, and a consequence of this has been a conversion of species-rich heathland to species-poor grasslands and forest. The mix of plant and animal species adapted to sandy, infertile soils is being lost because of nitrogen enrichment.

Increasing nitrogen in aquatic ecosystems can also have undesirable impacts. The Mississippi River drains nearly one-third of North America, and changes in water quality in the river over the last 50 years have triggered large ecosystem impacts in the northern part of the Gulf of Mexico. The problem is nitrogen in the water, and the principal cause is a dramatic increase in fertilizer nitrogen input into the Mississippi River drainage basin between the 1950s and 1980s. Since 1980, the Mississippi River has discharged, on average, about 1.6 million metric tons of total nitrogen to the gulf each year. Other nutrients, such as phosphorus, have not increased in the outflow. About 90% of the nitrate in the river comes from excess fertilizer draining off agricultural land and drainage from feedlots for cattle. The principal sources of nitrate are river basins that drain agricultural land in southern Minnesota, Iowa, Illinois, Indiana, and Ohio.

Nitrogen does not seem to be a primary limiting factor for algae in river systems; the ecological damage starts when these waters reach the coastal zone in the Gulf of Mexico. Coastal waters around the world are suffering from pollution—nutrients draining from the land and stimulating algal growth in the sea. In coastal waters off Louisiana, the excess nitrogen stimulates algal growth and associated zooplankton growth. Fecal pellets from zooplankton and dead algal cells sink to the bottom, and as this organic matter decomposes, the bacteria use all the oxygen in the bottom layer of water. At dissolved oxygen levels of less than 2 milligrams per liter all animals either leave or die. This shortage of oxygen in the bottom layer of coastal waters produces "dead zones."

Each summer, the Mississippi River outflow produces a dead zone in the northern Gulf of Mexico along the Louisiana-Texas coast that varies in size up to 20,000 square kilometers, the size of New Jersey (Bianchi et al. 2010). The hypoxic zone is most pronounced from June to August but can begin as early as April and last until October, when storms and winds mix up the surface and bottom water. Spawning grounds of fish and migratory routes of commercially harvested fish species are affected by the dead zones. To reduce the dead zones in the Gulf of Mexico, the most effective actions would be to reduce the amount of fertilizer usage and to keep the nitrogen in the agricultural fields with alternative cropping systems. The important message is that alleviating the problem of dead zones in the Gulf of Mexico requires an ecosystem approach to the whole catchment of the Mississippi River. An ecological understanding is needed of how nutrients put out in fertilizer to grow corn in Iowa can impact algal populations thousands of kilometers away in the Gulf of Mexico.

The nitrogen cycle has been heavily impacted by human activities during the last 50 years. It is urgent that national and international efforts be directed to reversing these changes and moderating the adverse impacts on ecosystems. A global program to investigate the consequences of human-induced changes in nutrient cycling is now being directed at the University of Minnesota (http://www.nutnet.umn.edu/home), and research is now under way to measure the detailed global impacts of humans on nutrient cycling. The most obvious direct impact on humans that flows from these changes in nutrient cycling are manifest now in global climate change, which we will discuss in Chapter 11.

ACID RAIN AND THE SULFUR CYCLE

Human activity through the combustion of fossil fuels has altered the sulfur cycle more than any of the other nutrient cycles. While human-produced emissions of nitrogen are only about 5 to 10% of the level of natural emissions, for sulfur we produce about 160% of the level of natural emissions. One clear manifestation of this alteration of the sulfur cycle is the widespread problem of acid rain in Europe and North America. Acid precipitation is defined as rain or snow that has a pH of less than 5.6. Low pH values are caused by strong acids (sulfuric acid, nitric acid) that originate as combustion products from fossil fuels.

Acid rain emerged as a major environmental problem in the 1960s, when damage to forests and lakes in Europe and eastern North America began to be noticed on a wide scale because oxides of sulfur and nitrogen could be carried hundreds of kilometers and then deposited in rain and snow. Lakes in eastern Canada were dying because of air pollution from the Midwestern states. Lakes in southern Norway were losing fish because of acid rain from England. By 1980 over large areas of Western Europe and eastern North America, annual pH values of precipitation averaged between 4.0 and 4.5, and individual storms produced acid rain of pH 2 to 3, the pH of vinegar.

Sulfur released into the atmosphere is oxidized to sulfate (SO_4) quickly and redeposited rapidly on land or in the oceans. Short-term events like volcanic eruptions contribute to the global sulfur cycle and make it difficult to estimate the equilibrium state of the atmosphere. Ore smelters and electrical generating plants have increased emissions during the past 100 years. To offset local pollution problems, smelters and generating plants have built taller stacks, which reduce pollution at ground level. Tall stacks (over 300 m) now are the standard, and they have exported the pollution problem downwind. Ice cores from Greenland show large increases in SO_4 deposition from the atmosphere in the last 60 years.

The United States and most developing countries have reduced sulfur dioxide emissions during the past 35 years. In the United States sulfur dioxide emissions were reduced 78% between 1980 and 2012. Reduced emissions have reduced the surface deposition of acid rain. But the effects of acid rain do not go away immediately as sulfur dioxide emissions fall, and the key question remains: *will forest and aquatic ecosystems recover from the effects of acid rain, and at what rate?* At Hubbard Brook the impact of acid rain has been to leach calcium from the soil to the extent that available calcium may limit forest growth, particularly for sensitive tree species like sugar maple. Stream water chemistry at Hubbard Brook is slowly recovering from acid rain, and at least another 10–20 years will be needed for streams to recover, even if sulfur dioxide emissions continue to decrease.

The clearest effects of acid precipitation have been on fish populations in Scandinavia and eastern Canada. Fish populations were reduced or eliminated in many thousands of lakes in southern Norway and Sweden once the pH in these waters fell below pH 5. In Canada, lakes containing lake trout have been the principal focus of research on the impacts of acid rain. Lake trout disappear in lakes once the pH falls below 5.4, and the cause is reproductive failure because newly hatched trout die. Lake trout are a keystone predator in many Canadian lakes, and they disappear slowly in lakes of low pH. Adult trout do not seem to be affected by low pH, nor is there any food shortage at low pH. The impact is on the small juveniles, and this causes a slow decline in the trout population over 10–20 years. Once lake trout are gone, acid-tolerant fishes like yellow perch and cisco become more abundant, and the food web loses species (Figure 9.5). Below pH 4 almost no fish can survive.

The changes that humans have made to the sulfur cycle have the potential to change nutrient cycling in natural ecosystems in a great variety of ways we cannot yet understand, much less predict. We cannot continue this aerial bombardment of ecosystems in the naive belief that nutrient cycles have infinite resilience to human inputs. Recent efforts to curtail sulfate emissions from fossil fuels have reduced the emissions of SO_4, and we must continue to press for further reductions. Once acidic precipitation is reduced, both forest and lake ecosystems can begin to recover from

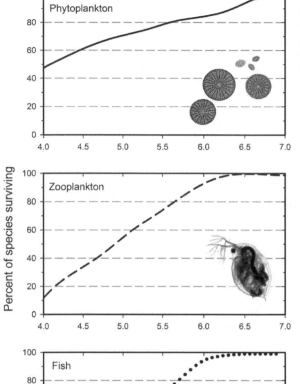

Figure 9.5 The effects of acid rain in eastern Canadian lakes on phytoplankton, zooplankton, and fish species surviving in the lakes. Once the pH of the lake water drops below about 6.5 species begin disappearing. (Data from Gunn and Mills 1998.)

the damage inflicted, but the time frame for recovery may be longer than we think.

NUTRIENT CYCLING IN PEATLANDS

Natural communities seem to exist in a state of long-term nutrient balance, but some exceptions stand out. Peatlands are areas in which plant production exceeds decomposition so that organic matter accumulates, and with it a deposit of nutrients. Peatlands cover only about 3% of the land surface of the Earth, but they store about 30% of the Earth's soil carbon. Most peatlands are in the Northern Hemisphere on areas covered by glaciers until about 10,000 years ago. Canada, Ireland, Scotland, Finland, and Siberia contain large areas of peatland. In peatlands, input exceeds output. Why do peatlands form, and how do nutrients cycle in peatlands? These questions are of increasing importance in the era of climate change because peatlands are a large storage system for carbon, and thus there is concern that global warming may cause this stored carbon to be released into the atmosphere as additional carbon dioxide.

Peat is produced in any area where moisture is in excess supply throughout the growing season so that the soil is saturated with water. Excessive moisture selects for certain kinds of plants because it retards decomposition so nutrients are not released into the soil. As leaves and stems die and fall to the soil surface they are only partly decomposed, and litter begins to build up. The critical boundary in a peat bog is the depth of the water table in summer. Above this boundary, the active zone, plant materials are broken down rapidly because oxygen is present. Below this boundary, the inactive zone, plant materials are broken down very slowly because no oxygen is present (microorganisms use it up more quickly than it can diffuse in).

Peatlands can be produced in any climatic zone, but they are most extensively distributed in cold, wet regions. Canada and the USSR contain about 75% of the world's peat. Decomposition is stopped by freezing winter temperatures, and drainage is impeded on many areas underlain with permafrost (permanently frozen subsoil). Peatlands can be extensive outside of arctic regions. Coal deposits laid down in the Carboniferous Period 400 million years ago are just extensive peat bogs developed in a warm climate with high rainfall where decomposition was hindered by high water tables.

Peat accumulation depends on the balance of the production of new

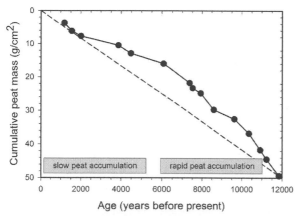

Figure 9.6 Peat accumulation pattern from the northeastern Tibetan Plateau, China for the past 12,000 years. The dashed line shows what would occur if the peat accumulation occurred at a constant rate. In this sedge-dominated peatland, peat accumulation occurred at a fast rate until about 6,000 years ago and then slowed. Dates were estimated using carbon-14. (Data from Hong et al. 2003.)

plant material and its decomposition. Peat accumulation varies with the climatic conditions of the area and the length of the growing season, but in most cases peat accumulates very slowly indeed. An average figure for northern temperate peatlands would be 20–80 cm of peat per 1,000 years (Wieder and Vitt 2006). As the climate changes, so does peat production. Peat was accumulating about 3 times as fast 6,000 years ago as it is today (Figure 9.6). Peat is an important fuel in Scotland, Ireland, and Russia, and is a renewable resource. But its rate of renewal is very slow and the rate of harvesting needs to be controlled.

Most peat bogs in the northern temperate zone seem to reach an equilibrium at 5–10 meters of depth in which the rate of addition of plant matter at the surface is balanced by decomposition at all depths. This equilibrium depth is dependent on drainage, vegetation, and temperature, and some peat deposits can grow very deep for periods of more than 50,000 years.

Nutrient cycling in peatlands is particularly interesting because of the reduced decomposition. Nitrogen and phosphorus are the two elements in shortest supply in peat bogs. There is an enormous capital of nitrogen in peat but most of it is completely locked up in complex organic compounds that plants cannot use for growth. Plants in peat bogs survive either by having a low nitrogen requirement and slow growth (as *Sphagnum* moss) or by utilizing nitrogen directly from the air. A few bog plants have root nodules containing bacteria that convert nitrogen from the air into nitrate, which the plant then uses for growth.

Phosphorus is very scarce in many peat bogs, and plants survive only by having low phosphorus requirements. There is no source for phosphorus except in rainwater or ground water, and the result is that when bogs are drained for agriculture few crops will grow without additional fertilizer being added because both phosphorus and nitrogen are in short supply. Given time and oxygen, peat will break down into humus, and consequently peat is widely used as organic manure to improve the structure of agricultural soils.

Even though peatlands on average have acted as carbon sinks, thus partly mitigating our increasing CO_2 emissions, they are at the same time the most important single methane source globally. Methane (CH_4) is produced by bacteria in the anaerobic lower layers of peatlands. Northern peatlands are suggested to contribute 34–60% of the global wetland methane emissions. Because methane has a global warming potential 23 times that of CO_2, boreal peatlands may contribute substantially to global warming over a 100-year horizon.

CONCLUSIONS

All life depends on nutrients, and understanding the cycling of nutrients is a key part of ecosystem studies. Some nutrients like oxygen are gaseous and move around the Earth on a large spatial scale, or they may be solid elements like calcium that have no means of long-distance transport. For any given ecosystem, the key point is that input must equal output or the ecosystem degrades. Inputs come from weathering of rock and soil, from the fixation of carbon, oxygen, and nitrogen by plants, or from volcanic activity. Outputs for gaseous elements can go into the atmosphere, or into storage sites in the bottom sediments of lakes and the ocean, or into peatlands in which the production of plant matter exceeds the rate of decomposition.

Nutrient cycles are critical for two reasons. If elements like phosphorus are in short supply and are required for crop plants, we need to close the nutrient cycle by recovering those elements we waste. If compounds like carbon dioxide are emitted in excessive amounts by the burning of fossil fuels, we must control their emission rate to stop runaway global warming caused by the increase in greenhouse gases. The simplest nutrient cycle is the water cycle, which we ignore at our peril because clean water is needed to protect human health. For these reasons nutrient cycles, which have often seemed to be esoteric events of no great importance to the human population, now have a central position in our developing understanding of how ecosystems operate sustainably.

CHAPTER 10

SOLAR ENERGY POWERS
NATURAL ECOSYSTEMS

KEY POINTS

- Every natural and agricultural ecosystem depends on energy input via solar radiation to drive plant photosynthesis.
- Only about 4–6% of solar energy can theoretically be converted into plant production via photosynthesis, and for most crop and natural ecosystems the efficiency of conversion is about 1%.
- Limitations on plant growth are set by a combination of light, water, temperature, and nutrients, and these factors define the Earth's capacity for plant production.

Much of ecology still deals with the details of particular species but in the last 20 years more and more research has focused on the ecosystem as a whole and concentrated on the physics of ecosystems as energy machines and nutrient processors. Exactly how plants and animals process energy and materials is important because it tells us about how ecosystems function and also because it has serious implications for how humans impact the Earth.

The metabolism of ecosystems can be understood as the sum of the metabolism of individual animals and plants. Individual organisms require a continual input of new energy to balance losses resulting from metabolism, growth, and reproduction. Individuals can be viewed as complex machines that process energy and materials. Organisms pick up energy and materials in two main ways. *Autotrophs* pick up energy from the sun and materials from nonliving sources. Green plants are autotrophs. *Heterotrophs* pick up energy and materials by eating living matter. Herbivores are heterotrophs that live by eating plants, and carnivores are heterotrophs that live by eating other heterotrophs. Communities are mixtures of autotrophs and heterotrophs. Energy and materials enter a biological community, are used by the individuals, and are transformed into biological structure only to be ultimately released again into the environment. The ecosystem level of integration includes both the organisms and their physical environment and is a comprehensive level at which to consider the movement of energy and materials.

Ecologists separate the chemistry and the physics of ecosystems by measuring two different processes.

1 *Flow of chemical materials.* We can view an ecosystem as a superorganism taking in food materials, using them, and passing them out. All chemical materials can be recycled many times through the community. A molecule of phosphorus may be taken up by a plant root, used in a leaf, eaten by a grasshopper that dies, and released by bacterial decomposition to reenter the soil. The process of the flow of chemicals was discussed in Chapter 9.

2 *Flow of energy.* We can view the ecosystem as an energy transformer that takes solar energy, fixes some of it in photosynthesis, and transfers this energy from green plants through herbivores to carnivores. Note that most energy flows through an ecosystem only once and is not recycled. Instead, it is transformed to heat and ultimately lost from the system. Only the continual input of new solar energy keeps the ecosystem operating.

Figure 10.1 illustrates the flows of energy and materials through the food web. Energy flow simplifies the great diversity of biological communities by reducing this diversity and all the species interactions to a single unit—the joule (or calorie).

PRIMARY PRODUCTION

Plants are the primary producers in an ecosystem. The process of photosynthesis is the cornerstone of all life on Earth. Photosynthesis is the process of transforming solar energy into chemical energy, taking up CO_2 from the air or water, converting it to carbohydrates, and in the process releasing oxygen into the air. Without photosynthesis none of us would be here, as there would be no oxygen on Earth. The simplest model of how ecosystems run would be to view light as the main driver of natural communities, since light is necessary for photosynthesis. The bulk of the Earth's living mantle is green plants (99.9% by weight); only a small fraction of life consists of animals (Whittaker 1975). How efficient are plants in capturing the energy in sunlight? How much plant production is limited by factors other than light? While light is critically important, it is not the only limiting factor for plant production, and it is important to look at the details of how different ecosystems operate.

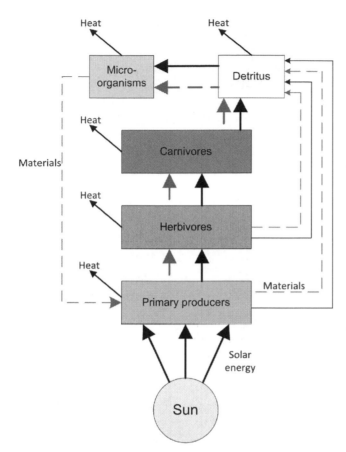

Figure 10.1 General representation of energy flows (dashed lines) and material cycles (solid lines) in the biosphere. Energy flows included are solar radiation, chemical energy transfers (in the ecological food web), and radiation of heat into space. Materials recycle from plants (primary producers) to herbivores and carnivores and eventually to detritus and back to the primary producers. (From DeAngelis 1992.)

EFFICIENCY OF PRIMARY PRODUCTION

How efficient is the vegetation of different communities as an energy converter? We can determine the efficiency of utilization of sunlight by the following ratio:

$$\text{Efficiency of primary production (\%)} = (100) \frac{\text{energy fixed by primary production}}{\text{total energy in incident sunlight.}}$$

The amount of solar radiation intercepted by the Earth is 21×10^{24} J per year, or about 8.1 J per cm^2 per minute. Plants utilize overall only about 0.02% of this total solar energy for photosynthesis (Figure 10.2), and most of the solar energy coming to the Earth is reflected back by the atmosphere or converted to heat. The energy in the sunlight arriving at the Earth's sur-

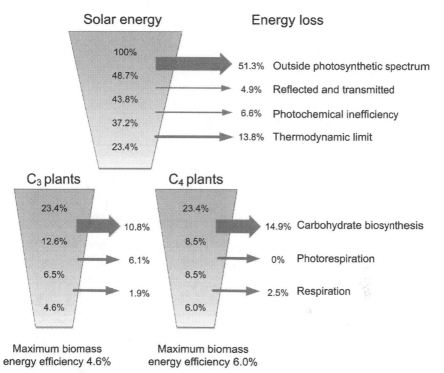

Figure 10.2 Solar energy losses for plant photosynthesis. The percentage of total full spectrum solar energy remaining (inside arrows) and percentage losses (at right) is shown from an original 100% arriving on Earth. Both C_3 plants (e.g., rice, wheat, coffee, potatoes) and C_4 plants (e.g., corn, sugar cane) are illustrated. The theoretical maximal photosynthetic energy conversion efficiency is 4.6% for C_3 plants and 6.0% for C_4 plants. (From Zu et al. 2010.)

face is reduced by more than half by atmospheric reflection or absorption, but still only a small amount of this incident energy is used in primary production. Gross primary production or carbohydrate synthesis is reduced in plants by respiration, the use of energy to maintain plant metabolism, so that net primary production, the energy that we can gain in crops or herbivores can eat, is reduced to a maximum of about 6% of the sun's total energy input to Earth.

The chemistry of photosynthesis is somewhat complicated because there are three different photosynthetic pathways that have evolved in plants: the C_3 pathway, the C_4 pathway, and the CAM pathway. All plants capture CO_2 out of the air to make carbohydrates, but they do it in different ways. C_3 plants capture CO_2 first in a 3-carbon compound in a slow reaction that does not operate well at low CO_2 levels. C_4 plants capture CO_2 in a 4-carbon compound in a faster reaction that operates even at low CO_2 levels, and in addition C_4 plants utilize sunlight more efficiently The CAM pathway is used by cacti and many desert succulents, and in this photosynthetic pathway CO_2 is picked up at night as a way of reducing water loss, and then the stored CO_2 inside the leaves is used when daylight arrives to complete photosynthesis. For our purposes many common temperate zone plants use the C_3 pathway, and C_4 plants are more common in tropical areas. Most of our crops like wheat and rice are C_3 plants, and a few crops like corn and sugar cane are C_4 plants.

The key ecological issue is that C_4 plants are more productive than C_3 plants because they achieve maximum photosynthesis at full sunlight. By contrast C_3 plants reach the maximum rate of photosynthesis at one-quarter to one-third of full sunlight. C_3 plants grow best at 20–25°C, while C_4 plants grow best at 30–35°C. C_3 plants increase photosynthesis in high-CO_2 environments, while C_4 plants are relatively unaffected by high CO_2 levels. The chemistry of these pathways is described in detail by Björkman and Berry (1973), Pearcy and Ehleringer (1984), and Blankenship (2002).

The bottom line is that plants can convert only about 4.5 to 6% of solar energy into net primary production. But this is a maximal rate, and plants in different ecosystems can rarely achieve this level of energy conversion.

Phytoplankton communities have very low efficiencies of net primary production, usually less than 0.5%, although rooted aquatic plants and algae in shallow waters can have higher efficiencies. The efficiency of gross primary production is higher in forests (2.0–3.5%) than in herbaceous communities (1.0–2.0%) or in crops (less than 1.0%) (Zhu et al. 2010, Blankenship et al. 2011). Forest communities are relatively efficient at capturing solar energy, but no natural vegetation type anywhere is able to capture more than 3–4% of the solar energy falling on the Earth. In forests, 50 to 75% of gross primary production is lost to respiration, so that net production may be only one-fourth that of gross production, much less than the theoretical maximum efficiency of a C_3 plants (Amthor and Baldocchi 2001). Forests have larger amounts of stems, branches, and roots to support than do herbs, and less energy is lost to respiration in herbaceous and crop communities (45–50%). The result of these losses is that for a broad range of terrestrial communities, about 1% of the sun's energy during the growing season is converted into net primary production.

If there is an upper limit to the efficiency of plants in capturing solar energy, it is important to ask whether it will be possible to increase the efficiency of agricultural crops to capture more solar energy. There is much research at present to select for crop plants that are more efficient at capturing the sun's energy, but this is not a simple problem. Algae may be a more efficient crop plant (Blankenship et al. 2011). Algal cultures under ideal conditions can achieve 5–7% energy efficiency, algal agriculture has been suggested as one way to provide more food for undernourished people as well as more energy through biofuels (Walker 2010, Blankenship et al. 2011). Molecular biologists are currently exploring ways of improving on the chemistry of plant photosynthesis to increase crop production (Zhu et al. 2010). Compared to standard silicon-based solar panels, which achieve a 10–20% conversion of solar energy into electricity, plants are inefficient.

PRIMARY PRODUCTION IN NATURAL ECOSYSTEMS: THE OCEANS

How does primary production vary over the different types of vegetation on Earth? Figure 10.3 illustrates the yearly production for ocean areas of the Earth. Productivity of the open ocean is very low, approximately the same as that of the arctic tundra. But since oceans occupy about 71% of the total surface of the Earth, total oceanic primary production adds up to

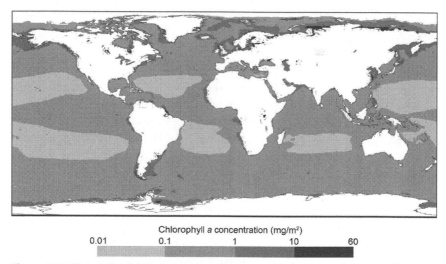

Chlorophyll a concentration (mg/m²)

0.01   0.1   1   10   60

Figure 10.3 Global distribution of marine primary production (annual averages) indexed by chlorophyll *a*. Ocean surface (to 30 meters depth) chlorophyll *a* concentration, 1997–2007 mean values from SeaWIFS satellite. Ocean chlorophyll *a* concentration is highly correlated with net primary production. (From NASA SeaWIFS, 2013.)

about 46% of the overall production of the globe. The surprise from these ocean data is that tropical oceans are very unproductive, and temperate and polar oceans are highly productive. If solar radiation alone controlled the rate of primary production, this pattern should be reversed. What ecological factors limit primary production in the oceans?

Light is the first variable one might expect to control primary production, and the depth to which light penetrates in an ocean is critical in defining the zone of primary production. Water absorbs solar radiation very readily. More than half of the solar radiation is absorbed in the first meter of water, including almost all the infrared energy. Even in "clear" water, only 5 to 10% of the radiation may be present at a depth of 20 meters. Very high light levels can inhibit photosynthesis of green plants, and this inhibition can be found in tropical and subtropical surface waters throughout the year. When surface radiation is excessive, the maximum in primary production will occur several meters beneath the surface of the sea.

If light is the primary variable limiting primary production in the ocean, there should be a gradient of increasing productivity from the poles toward the Equator. Figure 10.3 shows the opposite—there is no gradient of production from the poles to the Equator. Large parts of the tropics and subtropics, such as the Sargasso Sea, the Indian Ocean, and the Central Gyre of the North Pacific, are very unproductive. In contrast, the North Atlantic, the Gulf of Alaska, and the Southern Ocean off New Zealand are quite productive. The most productive areas are the western coastal areas of Africa, North America, and South America.

Why are tropical oceans unproductive when the light regime is good all year? *Nutrients* appear to be the primary limitation on primary production in the ocean. Two elements, nitrogen and phosphorus, often limit primary production in the oceans. One of the striking generalities of many parts of the oceans is the very low concentrations of nitrogen and phosphorus in the top 30 m of the ocean, where the phytoplankton live, whereas the deep ocean water contains much higher concentrations of nutrients.

The discovery that nitrogen limits primary production in many parts of the ocean was completely unexpected because nitrogen is abundant in the air and can be converted into a usable form by nitrogen-fixing cyanobacteria that are common in ocean waters. The expectation had been that phosphorus must be limiting productivity in the ocean because phosphorous does not occur in the air (except in dust). But this has turned out to be completely wrong, a good example of why "obvious" conclusions in science may not be correct. But the importance of nitrogen as a limiting factor raises another dilemma because several large parts of the oceans contain high amounts of nitrate and low amounts of phytoplankton. For example, the surface waters of the equatorial Pacific have both high nitrate and high phosphate concentrations but low algal biomass. One explanation for these oceanic regions is that they are communities dominated by herbivores that control plant biomass, and if plant biomass is low, nutrients are always in excess. An alternative explanation is that these parts of the ocean are limited by some nutrient other than nitrogen or phosphorous.

The Sargasso Sea is an area of very low productivity in the subtropical part of the Atlantic Ocean. The seawater there is among the most transparent in the world, and the surface waters are very low in nutrients. Nitrogen and phosphorus, however, do not seem to be limiting primary production, and iron seems to be critical. This was shown by a series of nutrient-enrichment experiments in which surface water from the Sargasso Sea was placed in bottles and enriched with various nutrients.

The demonstration of iron limitation in the Sargasso Sea stimulated the hypothesis that iron limitation could be responsible for the low productivity of the equatorial Pacific. Iron comes to the oceans largely as wind-blown dust from the land, and dust is particularly scarce in the Pacific Ocean and in the Southern Ocean. Iron is an essential component of the photosynthetic machinery of the cyanobacteria that fix nitrogen in the oceans. The impact of iron on primary production is mainly through its impact on nitrogen fixation, so that we have a sequence of potential limitations that operate in iron-poor parts of the ocean:

$$iron \rightarrow cyanobacteria \rightarrow nitrogen\ fixation \rightarrow phytoplankton.$$

There may be competition for iron from other kinds of bacteria in the ocean that are iron limited, reducing the amount going to cyanobacteria. In most of the open oceans, light is always available for photosynthesis, but nitrogen is not.

One controversy that arose from these findings about iron limitation was the suggestion that global rises in CO_2 levels could be stopped by fertilizing the unproductive oceans with small amounts of iron, thus stimulating ocean photosynthesis, which would take up superabundant CO_2. A series of detailed large-scale iron fertilization experiments in the Southern Ocean showed clearly that iron addition could stimulate photosynthesis there but that the effects were small on a global scale and thus not a poten-

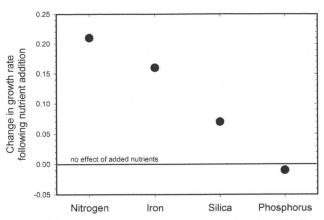

Figure 10.4 Effects of nutrient addition on marine phytoplankton growth rates in 303 experiments. Excess nutrients were added to a large ocean water sample and followed for 2–7 days. Nitrogen and iron are clearly the main limiting factors. Silica limitation occurs when diatoms are the dominant species in the phytoplankton. The black line marks the line of zero effect. Phosphorus never seems to be the limiting factor. (After Downing et al. 1999.)

tial vehicle for reducing CO_2 in the atmosphere (Buesseler et al. 2004, Boyd et al. 2007).

To quantify the relative effect of different limiting nutrients in the oceans Downing et al. (1999) analyzed 303 nutrient-addition and control experiments carried out over the last 40 years. They found that nitrogen addition stimulated phytoplankton growth most strongly, followed closely by iron addition (Figure 10.4). These results are consistent with the conclusion that nitrogen and iron are key limiting resources in the oceans.

Compared with the land, the ocean is very unproductive; the reason seems to be that fewer nutrients are available. Rich, fertile soil contains 5% organic matter and up to 0.5% nitrogen. One square meter of soil surface can support 50 kilograms dry weight of plant matter. In the ocean, by contrast, the richest water contains 0.00005% nitrogen, four orders of magnitude less than that of fertile farmland soil. One square meter of rich seawater could support no more than 5 grams dry weight of phytoplankton. In terms of standing crops, the sea is a desert compared with the land. And although the maximal rate of primary production in the sea may be the same as that on land, these high rates in the sea can be maintained only for a few days, unless upwelling enriches the nutrient content of the surface water.

Areas of upwelling in the ocean are exceptions to the general rule of nutrient limitation. The largest area of upwelling occurs in the Antarctic Ocean, where cold, nutrient-rich deep water comes to the surface along a broad zone near the Antarctic continent (Figure 10.3). Other areas of upwelling occur off the coasts of Peru and California, as well as in many coastal areas where a combination of wind and currents moves the surface water away and allows the cold deep water to move up to the surface. In these areas of upwelling, fishing is especially good, and in general there is a superabundance of nitrogen and phosphorus for the phytoplankton.

One of the most exciting recent developments in marine ecology is the ability to estimate primary production from satellite remote sensing data. Chlorophyll concentration in the surface water can be estimated by spectral reflectance using blue/green ratios. Remote sensing by satellites allows marine ecologists to analyze large-scale production changes in the ocean without being limited to a few measurements made off a ship. These techniques promise to enlarge our understanding of primary production in the oceans and how it varies in space and time.

In summary, total primary production in the ocean is rarely limited by light but by the shortage of nutrients, particularly nitrogen and iron, which are critical for plant growth. Phosphorus limitation of primary production is rare in oceanic ecosystems. Solar power is essential for oceanic primary production but it is not the limiting factor.

PRIMARY PRODUCTION IN FRESHWATER LAKES AND STREAMS

In freshwater communities, the same limiting factors that operate in the ocean do not seem to operate. Solar radiation limits primary production on a day-to-day basis in lakes, and within a given lake you can predict the daily primary productivity from the solar radiation (Horne and Goldman 1994). Temperature is closely linked with light intensity in aquatic systems and is difficult to evaluate as a separate factor. Everything else being equal, warmer lakes will be more productive than colder lakes. Given light and reasonable temperatures, nutrient limitations operate to control primary production in freshwater lakes, and the great variety of lakes is associated with a variety of potential limiting nutrients. For growth, plants require nitrogen, calcium, phosphorus, potassium, sulfur, chlorine, sodium, magnesium, iron, manganese, copper, iodine, cobalt, zinc, boron, vanadium, and molybdenum. In addition to light and temperature, the major limiting factors in freshwater are most likely to be the macronutrients nitrogen, carbon, and phosphorus.

During the 1970s the problem of what controls primary production in

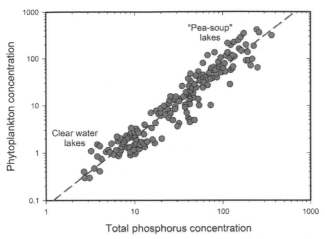

Figure 10.5. Freshwater lakes: the relationship between phosphorus concentration (grams per liter) and summer phytoplankton standing crop (measured by chlorophyll as grams per liter). Phosphorous concentration varies 100-fold in these lakes and algal concentration in the water varies 1000-fold. Phosphorus levels limit primary production in many freshwater ecosystems. (Data from Ahlgren et al. 1988.)

freshwater lakes became acute because of increasing pollution. Nutrients added to lakes directly in sewage or indirectly as runoff had increased algal concentrations and had shifted many lakes from clear-water lakes dominated by diatoms or green algae to green, pea-soup lakes dominated by blue-green algae. This process of lake pollution and transformation is called *eutrophication*. Before we can control eutrophication in lakes, we have to decide which nutrients need to be controlled. Three major nutrients were suggested—nitrogen, phosphorus, and carbon—and much experimental work was carried out from 1960 to 1985 to determine the major limiting factor. The conclusion was simple: phosphorus is the limiting nutrient for phytoplankton production in the majority of freshwater lakes (Edmondson 1991). The standing crop of phytoplankton in a lake is highly correlated with the total amount of phosphorus in the lake water (Figure 10.5).

The practical advice that has followed from these and other experiments is to control phosphorus input to lakes and rivers as a simple means of checking undesirable algal growth. One result of this research was the production of phosphorus-free detergents for laundry. A second useful result was that the amount of phosphorus that can be added to a lake can be calculated so that planners can design human land use around any lake in a way that maintains high water quality.

Estuaries are mixtures of freshwater and saltwater, and are often heavily polluted with nutrients from sewage and industrial wastes. Because they form an interface between saltwater, in which nitrogen is often limiting to phytoplankton, and freshwater, in which phosphorus is typically limiting, estuaries contain complex gradients of nutrient limitation in which added phosphorus and nitrogen from pollution can strongly affect primary production (Doering et al. 1995).

To summarize, in freshwater communities, primary production is usually limited by light, temperature, and phosphorus levels.

PRIMARY PRODUCTION IN TERRESTRIAL COMMUNITIES

In terrestrial habitats, temperature ranges are much greater than in aquatic habitats, and the great variation in temperature from coastal to alpine or continental areas makes it possible to uncouple the solar radiation–temperature variable, which is so tightly bound in aquatic systems. The large seasonal changes in radiation and temperature are reflected in the global patterns of primary production. Using satellite imagery, we can now look at continental and global patterns of terrestrial productivity. Satellites, such as the NOAA (National Oceanic and Atmospheric Administration of the USA) series of meteorological satellites and the NASA SeaStar spacecraft, have sensors on board that record spectral reflectance in the visible and infrared regions of the electromagnetic spectrum. As green plants photosynthesize, they display a unique spectral reflectance pattern in the visible (0.4–0.7 µm) and the near-infrared (0.725–1.1 µm) wavelengths (Goward et al. 1985). Vegetation indices, which discriminate living vegetation from the surrounding rock, soil, or water, have been developed by combining these spectral bands. The AVHRR (advanced very high resolution radiometer) sensor on the NOAA satellite and the Sea WiFS (Sea-viewing wide field-of-view sensor) on the SeaStar spacecraft are especially useful for monitoring global vegetation because they have worldwide coverage at a resolution of 1.1 kilometers at least once per day in daylight hours (Signorini et al. 1999).

The availability of satellite data to estimate primary production on a global basis has made it relatively easy to obtain measurement data of primary production of terrestrial communities. Table 10.1 brings together the data on primary production for all terrestrial ecosystems. On land pri-

Table 10.1 Net primary production and estimated biomass for the major ecosystems of the Earth. (Data from Schlesinger and Bernhardt 2013, table 5.3.)

Ecosystem	Area (10⁶ km²)	Net primary production (g C per m² per year)	Total net primary production (10¹⁵ g C per year)	Biomass (g C per m²)
Tropical forests	17.5	1250	20.6	19,400
Temperate forests	10.4	775	7.6	13,350
Boreal forests	13.7	190	2.4	4150
Mediterranean shrublands	2.8	500	1.3	6000
Tropical savannas and grasslands	27.6	540	14.0	2850
Temperate grasslands	15.0	375	5.3	375
Deserts	27.7	125	3.3	350
Arctic tundra	5.6	90	0.5	325
Crops	13.5	305	3.9	305
Ice	15.5	0	0	0
Total	149.3		58.9	

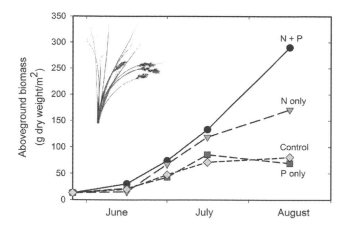

Figure 10.6 Effects of fertilization with nitrogen and phosphorus on primary production in a subarctic salt-marsh dominated by the sedge *Carex subspathacea*, southern Hudson Bay, Canada. Nitrogen and phosphorus were the primary limiting factors to growth. (From Cargill and Jefferies 1984.)

mary production is limited by light, temperature, water, and nutrients in the soil. Tropical ecosystems are the most productive on an annual basis, and crops represent only a small fraction of the Earth's total primary production.

Within the climatic constraints dictated by light, temperature, and rainfall, soil nutrients limit production. Farmers and fertilizer companies have known this for many years. Ecologists wish to know exactly which nutrients are limiting and exactly how much primary production can be stimulated by adding nutrients. Nutrient-addition experiments on local sites can be used to determine how much primary production is limited by nutrients. Cargill and Jefferies (1984) added nitrate and phosphate to salt-marsh sedges and grasses in the subarctic to test for nutrient limitation. Figure 10.6 shows that in the absence of grazing, the addition of nitrate doubled primary production of the sedges and grasses, and the joint addition of phosphate and nitrate quadrupled production. In this marsh, as in many terrestrial communities, nitrogen is the major nutrient limiting productivity, and when nitrogen is suitably increased, phosphorus becomes limiting. A sequence of limiting factors is a key concept for understanding different levels of primary production in human-managed agricultural and forestry ecosystems.

From a global perspective primary production is largely driven by the physical environment in the form of light, temperature, rainfall, and nutrient supplies. Plants have adapted to these constraints from the environment to produce through photosynthesis the materials that drive all the subsequent biota, including ourselves.

CONCLUSIONS

Incoming solar radiation drives all of our natural and agricultural ecosystems via the process of photosynthesis, in which plants convert CO_2 into carbohydrates and release oxygen. Plants capture at most about 6% of solar energy, and after the limitations imposed by temperature, water, and nutrients the efficiency of use of solar energy by plants is only about 1%. Ocean productivity is very low per unit of area, and nitrogen and iron are often limiting factors on primary production. Tropical seas have clearwater because they are very unproductive. By contrast, freshwater systems are typically limited by phosphorus, and the addition of phosphorus in sewage has greatly increased freshwater pollution around cities. Terrestrial ecosystems are limited by temperature, water, and nutrients, particularly nitrogen but also by phosphorus and the micronutrients that come from the soil. Increasing agricultural production involves analyzing all of these limitations, from the efficiency of light capture by leaves to the nutrient balance of the soil, and is now a critical issue for the growing human population.

CHAPTER 11

CLIMATES CHANGE, COMMUNITIES AND ECOSYSTEMS CHANGE

KEY POINTS

- There is no question that the Earth's climate is changing rapidly and that the major cause of the observed warming is the accumulation of greenhouse gases in the atmosphere from our carbon economy. Carbon dioxide and methane are the major causes of global warming.
- Biological indicators of climate change back up the limited temperature and rainfall records of the last 200 years. Tree rings, lake and ocean cores, and coral growth rings all show how highly variable climate has been for the last 30,000 years.
- Biological communities change slowly over hundreds to thousands of years in response to climatic variation. Slow biological adaptation in the past contrasts markedly with our current very rapid shifts in climate driven by greenhouse gases.

We are living in the era of rapid climate change, and it is useful to look back on the Earth's history to see how climate change has affected ecosystems and to get a time perspective on the rates of changes that have occurred in the past. Evidence for changes in climate comes from a variety of sources. We have direct measurements of temperature changes only for the past 100–200 years, but we can infer much longer-term changes from indirect measures of temperature in tree rings, coral skeletons, and ice cores.

PHYSICAL RECORDS OF CHANGING CLIMATE

Most of the long-term records of climate change rely on the physics of the greenhouse effect, one of the most well established theories of atmospheric physics and chemistry. The greenhouse effect arises because the Earth's atmosphere traps heat near the surface. Water vapor, CO_2, and trace gases absorb the longer infrared wavelengths emitted by the Earth. An increase in the concentration of greenhouse gases thus tends to warm the Earth by reradiation. None of this is controversial, and the greenhouse

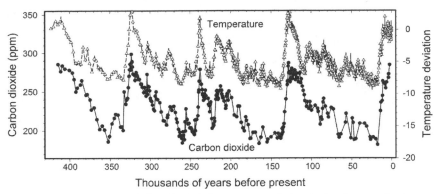

Figure 11.1 Long-term variations in global temperature (triangles) and atmospheric carbon dioxide concentration (circles) determined from the 3,623 meter–long Vostok ice core, Antarctica, covering the last 420,000 years. Carbon dioxide can be measured from air trapped in the ice as it is formed. Temperature is measured as the deviation from present day temperatures (°C). The temperature changes show the last 4 ice ages, each about 100,000 years long. There is a high correlation between CO_2 levels and global temperature. (Modified after Petit et al. 1999, Intergovernmental Panel on Climate Change 2013.)

principles apply as well on Earth as they do on Venus (dense CO_2 atmosphere, very hot) and Mars (thin CO_2 atmosphere, very cold).

Ice cores have provided one way of measuring climatic changes. Air is trapped by snow as it is transformed into glacial ice, and by taking ice cores one can sample the atmosphere back in time. The most spectacular example of this method is a 3,623 m–long ice core collected by the Soviet Antarctic Expedition at Vostok, Antarctica (Petit et al. 1999). This ice core spans 420,000 years. Temperature at the time of ice formation can be determined by the ratio of oxygen-18 to oxygen-16 in the ice, and CO_2 levels can be determined from the air trapped in the ice. The resulting time series of changes in the Vostok ice core are shown in Figure 11.1. There is a close correlation between carbon dioxide in the air and global temperatures over the past 420,000 years.

There are two difficulties in extrapolating these past correlations between CO_2 levels and temperature into the future. First, our present CO_2 levels of 400 ppm exceed the levels of 280 ppm in 1750 at the start of the Industrial Revolution, and levels found in nature during the last 420,000 years. We do not know if we can extrapolate from the relationships found in the past. Second, humans are changing CO_2 levels very rapidly from year to year, whereas the historical changes in CO_2 were relatively slow. We do

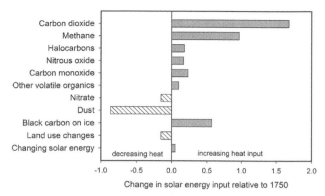

Figure 11.2 Changes in radiative heating of Earth from the base level in 1750 to 2011. Increases in carbon dioxide and methane are the main contributors to increased temperatures on Earth. Dust in the air is the main factor reducing incoming solar energy over this time period. (Data from IPCC 2013.)

not know the rates at which an equilibrium can be established between CO_2 sources and sinks in the ocean or on land

Other greenhouse gases can contribute to global warming as well as CO_2. The most important of these are methane, nitrous oxides, ozone, and chlorofluorocarbons (CFCs), and these trace greenhouse gases may together be as important or more important than CO_2 in the greenhouse effect of the 21st century. The atmospheric concentration of methane has increased about 150% since 1750 and is far above the average levels of the last 420,000 years. Fossil fuels, agriculture, and cattle production are major components of methane increases.

The major question the Intergovernmental Panel on Climate Change (IPCC) has been addressing for the past 25 years is why average global temperature has been increasing, particularly during the last 30 years. The task of estimating the effects contributed by all the culprits is difficult and there is always some error, even if it is all based on the laws of physics that almost everyone accepts. Figure 11.2 provides a summary of the impact of how much the different changes in the chemistry of the atmosphere have contributed to global warming. The two strongest contributors to climatic warming since 1750 are increased emissions of CO_2 and methane. Only one major contributor actually reduced incoming radiation and that was dust added to the atmosphere from soil erosion associated with agriculture and droughts.

Most people accept the findings summarized in Figure 11.2, and note that both CO_2 and methane are products of the Industrial Revolution, which began around 1750. If the physics is clear, the solutions to the problem are most difficult because they do not involve physics but social science and human behavior.

BIOLOGICAL RECORDS OF CHANGING CLIMATE

Most of the human impacts on ecological communities that we have described already in this book have been due to direct human actions like introducing new species that became pests (as illustrated in Figure 3.4) or removing species that are critically important for community structure (as illustrated in Figure 7.7). The longer-term impact involves the direct effects of climate change on communities. How have communities in the past responded to a changing climate?

The impacts of climate on biological systems can be traced in relatively simple systems that respond to changes in temperature. Tree growth rings record temperature and rainfall changes because ring width is wider in good years for growth and narrower in poor years. The differing ring widths affect the density of the wood deposited. This record of poor and good years for tree growth can be extended back as far as 1,500 years by cross dating (Figure 11.3). The most extensive chronology is the compilation of Scots pine (*Pinus sylvestris*) tree-ring data from northern Sweden, which has the longest continuous tree-ring width chronologies in the world, covering the period 5407 BC to AD 1997 (Grudd 2008). This 7,400-year precisely dated record was amalgamated from 880 individual

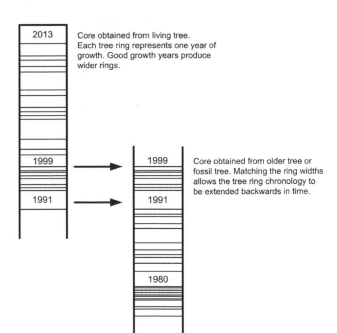

Figure 11.3 Tree ring analysis for dating. The diagram shows how cross-dating of tree rings is done and how a dated tree-ring chronology can be extended back in time. Such chronologies can now be extended back more than 1,000 years. (Modified after Fritts 1976.)

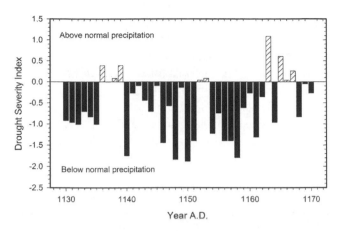

Figure 11.4 An example of a mega-drought in western North America from 1130 to 1170 AD. Drought was inferred from tree-ring widths covering the last 1000 years. Similar mega-droughts extending 20–40 years occurred in western North America four times in the medieval period from 1000 AD to 1400 AD. (Data from Herweijer et al. 2007.)

tree samples, collected from living trees and remnant wood preserved on dry ground and in small lakes. Tree growth in the north of Fennoscandia has a strong correlation with summer temperature, and tree-ring width is correlated with a short peak period of summer warmth. These tree ring data from Sweden show that the late 20th century was not exceptionally warm. On 10-year to 100-year timescales, periods around AD 750, 1000, 1400, and 1750 were equally warm, or warmer than the last 30 years. The 200-year-long warm period centered on AD 1000 was significantly warmer than the late 20th century. Climate in the past 1,500 years was highly variable.

In dry areas like the American Southwest, drought is the best correlate of tree-ring widths, and this has allowed ecologists to construct drought chronologies going back 1,000 years from the rings of drought-sensitive tree species. Drought is a major problem affecting agricultural production at the present time, and has been a major climatic disaster affecting humans throughout history and prehistory. These tree-ring reconstructions allow us to put recent droughts such as the 1930s Dust Bowl into the context of 1,000 year natural drought variability in North America. These tree-ring data have revealed the existence of successive "mega-droughts" of extreme persistence (20–40 years), yet similar to the major droughts experienced in today's North America (Herweiger et al. 2007). Figure 11.4

illustrates one of these mega-droughts from AD 1130 to 1170. These mega-droughts occurred during a 400-year-long period in the Middle Ages with a climate at that time varying as much as our current climate. The implication is that the mechanisms forcing persistent drought in the American West at the present time are similar to those underlying the mega-droughts of the Middle Ages.

The influence of changing climate on ecosystems can be seen clearly in fossil pollen. Pollen grains from plants are preserved as fossils in lake sediments. Most plant species have very specific pollen characteristics. Cores taken from lakes or swamps can be radiocarbon dated, and with pollen samples identified, we can obtain a record of vegetation changes over the past 30,000 years and correlate these vegetation changes with climatic change. Lucas and Lacourse (2013) analyzed a 9-meter core from a small lake on Pender Island in southwestern British Columbia to help define the history of forest changes in this area of the Pacific Coast of North America over the last 10,000 years since the melting of the ice sheets that blanketed the Northern Hemisphere during the last ice age. Figure 11.5 shows the pollen profile for this small lake. The pollen profile shows an herb-dominated community about 10,000 years before the present time, giving way over the next 2,000 years to alder and then Douglas fir and lodgepole pine forest. About 7,500 years ago Garry oak and big-leaf maple were relatively common, but after about 5,000 years ago Garry oak began to decline to a low level. At the same time, Douglas fir continued to increase and became the dominant tree in this region. Ferns that were common in the earlier years almost completely disappeared in the pollen profile after 7,500 years ago. The most noticeable changes in the vegetation on Pender Island during the last 10,000 years are the rise and decline of a Garry oak community between approximately 7,600 and 5,500 years ago, and the eventual dominance of Douglas fir (Lucas and Lacourse 2013).

The whole issue of community changes under climate change places a challenge on current conservation and restoration efforts. For the Garry Oak ecosystem, what should land managers and conservation ecologists manage to protect? If we select the profile at the end of the ice age 8,000–10,000 years ago, we have the state of this community before human occupation of the land. If we let plant succession take its present course, Garry oak will nearly disappear into a Douglas fir forest and this ecosystem will be lost. Restoration goals become more difficult when we take a changing community—in particular one with human-induced changes in the landscape—as the norm rather than a static community (McCune et al. 2013).

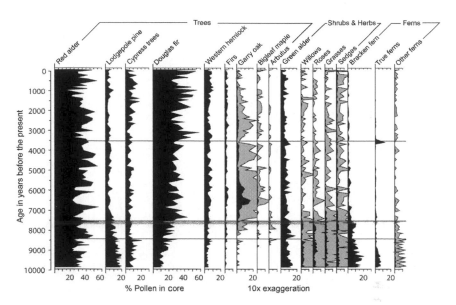

Figure 11.5 Pollen and spore percentages of plants for Roe Lake, Pender Island, southwestern British Columbia. Note that the plants with grey shading on the right side of the diagram have a 10× exaggeration applied to their pollen or spore frequency so as to make them visible on the profile. The hatched line across the diagram at 7,600 years ago denotes the position in this 9-meter core of the Mazama volcanic eruption deposits. Mount Mazama was a volcano in Oregon that destroyed itself in a large eruption and left as its remains Crater Lake, Oregon. (From Lucas and Lacourse 2013.)

Another record of climatic effects on biological systems comes from hard corals. The tropical oceans leave a climate record in living corals as well as in cores taken in the seabed that contain microfossils. Most reef corals live at depths of less than 20 meters and grow continuously at rates of 6–20 millimeters per year. Many hard coral species produce annual density bands (similar to tree rings) that can be seen in X-rays or under ultraviolet light. The skeletons of reef-building corals carry a diverse suite of isotopic and chemical indicators that track water temperature and salinity. For example, the ratio of strontium to calcium in coral cores traces sea surface temperatures at a time scale of three weeks, so that very fine temperature changes from the past can be measured. Many of these coral records have clearly defined annual chronologies extending back at least to the 15th century; when combined with seabed cores and terrestrial pollen, data can extend records of paleoclimates back 30,000 years (Reeves et al. 2013). With data from tree rings, lake and seabed cores, peat bogs, and

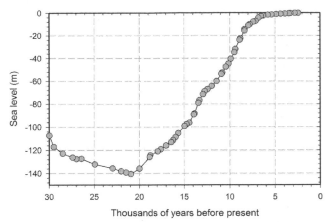

Figure 11.6 Sea level changes in Northern Australia and the Indonesian Archipelago for the past 30,000 years. The current position of sea level is set at zero, and the drop in sea levels in this tropical area during the last ice age exceed 130 meters. The study zone is bounded by 23.5°N, 160°E, 23.5°S, and 95°E). (From Reeves et al. 2013.)

other biophysical archives, ecologists can construct a global picture of how climate has varied globally and locally.

The tropical Australasian region illustrates how climate change in the past has affected tropical communities. Figure 11.6 shows the extent of changes in sea level since the peak of the last ice age, when a substantial amount of the Earth's water was stored on land in glaciers. Coastal plant communities would have been strongly affected by these changes in sea level. Figure 11.7 shows a composite of marine and terrestrial pollen core data and illustrates the changes in tropical vegetation associated with the end of the last glaciation and the subsequent changes in sea levels. The key point is that tropical vegetation is not immune to significant shifts in climate.

CONSERVATION ISSUES WITH CHANGING CLIMATE

Vegetation change accompanies climate change in the long term, and this raises an interesting question. Conservation biologists often wish to protect, restore, and conserve existing vegetation types for future generations. These goals are often defined by a historical baseline that occurred prior to extensive human disturbance, such as European settlement in Australia and North America. But if ecological communities were heavily influenced by native peoples prior to European settlement, conservation efforts may require more active management rather than simple removal

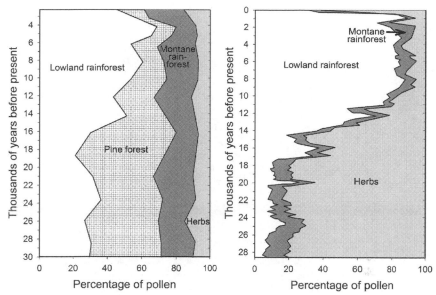

Figure 11.7 Pollen diagrams for two areas in Tropical Australia during the last 30,000 years. The marine core Bar 94-25 (left) was taken 100 kilometers off West Java in the Andaman Sea (6.43°S, 95.32°E). The West Java site Rawa Danau (right) is a terrestrial tropical lowland swamp on West Java (6.18°S, 105.95°E). Both of these cores show an increase in lowland rainforest as the ice age was closing about 10,000 to 15,000 years ago. (Modified from Reeves et al. 2013.)

of recent disturbances. The hidden problem is gradual changes in vegetation communities that are designated for conservation. A good example of this problem is the Garry oak savanna plant community of southwestern British Columbia (Bjorkman and Vellend 2010).

The dominant vegetation of southeastern Vancouver Island and the adjacent Gulf Islands of British Columbia is at present closed-canopy Douglas fir (*Pseudotsuga menziesii*) forest (cf. Figure 11.5), but a fragmented network of savanna habitat patches within the forest matrix harbors a rich diversity of native herbaceous plants, including over 100 threatened species. The savanna trees are largely Garry oak (*Quercus garryana*), and this species is a flagship for regional conservation efforts (Figure 11.8). Because savannas are open habitats, they are often designated for housing developments; this contributes to habitat loss. This part of western Canada was settled by Europeans in the mid-1800s, but native people have occupied the area for thousands of years. Pollen data suggest Garry oak savannas began to decline about 2,500 years ago, and historical data suggest the

Figure 11.8 The Garry oak woodland on the south end of Vancouver Island. This endangered plant community was maintained by fires set by Native Americans in the past, but is now undergoing change toward a closed forest of Douglas fir trees because fire has been eliminated in areas close to human settlements. It provides a conservation dilemma because it harbors many threatened plant species that cannot live in closed forest. Without fire or human management, this plant community will disappear.

hypothesis that the cultural practices and use of fire by indigenous peoples prior to European settlement heavily influenced the vegetative structure of this landscape and perpetuated the Garry oak savannas when otherwise they may have disappeared. This idea is supported by the historical data that show that fire-sensitive red cedar trees have been increasing during the last century while fire-resistant trees like Douglas fir have been declining. Fire is not the only influence on these savanna ecosystems, but lack of fires appears to be a dominant force driving recent community changes.

Restoration efforts are often prone to uncertainty about target conditions (Hobbs and Cramer 2008), especially in areas with no appropriate reference sites to help define historical conditions. Land managers often follow a do-nothing approach and allow land to return to its "natural" state. But in this case the open character of the endangered savannas on Vancouver Island was likely maintained by fires purposefully set by native peoples. Thus, restoration of these habitats to their pre-European state cannot be accomplished passively. Maintaining open savannas in this region would need to involve active management to remove encroaching trees and shrubs, either by controlled burning or mechanically by mowing. Furthermore, the tree species currently most closely associated with the savannas at the present time (Garry oak) does not appear historically to be a necessary dominant tree, since many open areas that were surveyed in the 1800s were populated with sparse Douglas fir trees. This highlights the need to understand community and ecosystem processes, and emphasizes the need for active management to achieve conservation and restoration goals.

HUMAN SOCIETIES AND CHANGING CLIMATE

There is much discussion and disagreement about the role of climate change as a major factor in the collapse of human societies in the past, stimulated by the book *Collapse* by Jared Diamond (2005). There are extensive data from ancient historical eras and extensive archaeological work on sites that in the distant past contained large settlements. As one would expect, a variety of causes can be postulated for the collapse of a human society from war, to disease, to migration, or to a loss of the plant or animal resource base on which the society relied.

A classic example from North America is the disappearance of large settlements of Native Americans from what is now the central United States after AD 1150 (Benson et al. 2009). Large settlements like Cahokia near St. Louis may have contained 10,000 to 15,000 people or more at their peak. From AD 1050 to 1100, the area along the Mississippi River was transformed by the construction of Cahokia, and a large influx of people moved there from small villages. Farming along the Mississippi River floodplain sustained the large settlement. Tree ring–based records of climate change in the area show that this rapid development occurred during one of the wettest 50-year periods of the last millennium. But for the next 150 years, a series of persistent droughts occurred in the Cahokia area. By about AD 1200, Cahokia's population had decreased by 50 percent, and by AD 1350, Cahokia and much of the central Mississippi valley had been abandoned. These were largely agricultural communities living on crops grown along the river floodplains, and drought is commonly suspected as a major cause of this collapse (cf. Figure 11.4), but social factors and migration could also be involved. What is clear is that by the time of European settlement in the 17th century, there were very few Native Americans in sites that previously had relatively large populations.

Another well-known collapse of human societies occurred in the Maya settlements of Central America from about AD 750 to 1050 (Middleton 2012). Millions of people died because of agricultural failure caused by a series of unusually severe and prolonged mega-droughts, as illustrated in Figure 11.4. The Maya collapse did not occur instantly, and there were periods between the droughts when rainfall was abundant. The major Maya collapse took place in four separate phases of abandonment (AD 760, 810, 860, and 910), and it has been postulated that the collapse was unavoidable and due directly to climatic factors reducing agricultural production.

It is always difficult to explain past events in human societies. In addition to climate change, social factors no doubt played an important role. Over the large area of the Maya civilization in Central America and over several centuries of change, there were certainly reversals of climatic conditions, so a single account to explain changes over a wide spatial and chronological area is always difficult. Rather than being a sudden collapse, the demise of the Maya settlements took well over a century to play out, and endemic warfare between competing sites contributed to the region's collapse. Small Maya states continued to exist along with the Spanish until the last was conquered in AD 1697.

Human societies are complex; consequently, many factors can be invoked to explain population responses to changing climate in the absence of detailed data. What is clear to an ecologist is that all human societies depend on food supplies from plants and animals, and the food production system depends on a continued, uninterrupted supply of water. Drought has been a major catastrophe for all human societies, and the absence of prolonged droughts is one factor that seems key to continued stable settlements.

CONCLUSIONS

Climate is the overall controller of the ecosystems we see on Earth today, so it is not surprising that climate change is a key focus for ecological research. The physical laws that allow the Earth to be an inhabitable planet are well known, and the key to climate lies in the greenhouse gases that surround us. Carbon dioxide and methane are the key drivers of current climate change, and the drastic increase of both of these is a result of our use of fossil fuels. Continued increase in these gases can be viewed only as catastrophic not only for the plants and animals of Earth but also for human societies.

Many biological indicators track changes in climate, and data from tree rings, core samples from the ocean and lakes, and coral reef expansion and contraction all provide us with a detailed description of how plant communities have changed in the past, particularly since the last glaciation ended

10,000 years ago. These biological data also give us a record of how quickly biological ecosystems can respond to climatic variation. The slow rates of climate change in past millennia contrast markedly with the rapid rates of change we are causing with the carbon economy.

Human societies can also be strongly affected by climate change, as we can see if we look at the historical and archaeological data that have been accumulated in the last 80 years. We assume our society will be the exception, but the key to many collapses in past human societies has been resources, particularly food production, which depends on a continuous supply of water. Drought driven by climate change is the number-one threat to our long-term survival.

CHAPTER 12

EXTINCTION IS FOREVER AND SPECIES LOSSES CAUSED BY HUMANS ARE AVOIDABLE

KEY POINTS

- Extinction is common in the long evolutionary history of life on Earth, but at the present time many species are being recognized as threatened because of human activities.
- Habitat loss and degradation are a major threat to many endangered species, followed closely by invasive pests and overharvesting by hunting or fishing. In general, the larger the size of a species, the higher the extinction risk.
- National parks and protected areas are essential for conservation but are rarely large enough to ensure species viability. Conservation on surrounding private lands is essential.
- Small populations are particularly susceptible to genetic decay from inbreeding and loss of genetic variability. Management is critical to maintaining genetic variation in small populations.

Dinosaurs came and went, and we go to museums to wonder at the size and variety of the skeletons that have been uncovered. The dinosaurs disappeared in a massive extinction event about 66 million years ago, when it is thought a comet or large meteorite hit the Earth and destroyed about 75% of the planet's existing life. These kinds of catastrophic events are the realm of geologists, and we assume that such extinctions could not occur at the present time. But we are possibly entering into another major extinction event in the history of life on Earth, in this case driven not by asteroids but by human impacts on the Earth's ecosystems. Since we control the present threats to species, we can do something about this problem, and understanding the threats to species today is a start in that direction. There are four categories of threats to threatened or endangered species: overharvesting, habitat loss, pest introductions, and food web collapse.

OVERHARVESTING

Overkill consists of fishing or hunting at a rate that exceeds a population's capacity to rebound. The species that are most susceptible to overkill are the large species with low reproductive rates, such as elephants, whales, and rhinoceros. Species on small islands are also vulnerable to extinction. The great auk, a large flightless seabird, was hunted to extinction on islands in the Atlantic Ocean in the 1840s because of a demand for feathers, eggs, and meat (Montevecchi and Kirk 1996).

The decline of the African elephant is a classic example of the effect of hunting on a large mammal. The African elephant is the largest living terrestrial mammal, weighing up to 7,500 kilograms. Sexual maturity is reached only after 10–11 years, and a single calf is born every 3 to 9 years. The potential rate of increase is about 6% per year, a low population growth rate. Between 1970 and 1989 half of Africa's elephants were killed for the ivory trade. This decline prompted the Convention on International Trade in Endangered Species (CITES) to ban all trade in ivory, and the response has been a dramatic increase in elephant numbers (Blake et al. 2007). From 2002 to 2006 elephant numbers in Africa increased on average 4% per year. Southern Africa holds about 58% of the continent's elephants, while east Africa holds 30%. The situation in central and west Africa is less clear because of little data (Blanc et al. 2007). In some of the national parks in southern Africa, elephants are considered overabundant and must be culled. The key to the extensive decline of elephants before 1990 was clearly poaching for ivory, and once this incentive was removed, populations began to recover. Unfortunately, by 2007 poaching on elephants for ivory began on a large scale again in central Africa, and the economics of the ivory trade began to threaten many populations (Dublin 2013).

Poaching or excessive harvesting is often thought to be the direct cause of decline in elephant population numbers, but as we gain understanding of social behavior we can begin to recognize social and genetic consequences that affect the ability of populations to recover. Archie and Chiyo (2012) describe these consequences for the African elephant. In long-lived animals, the social bonds between related individuals can be important for population health. African elephant females maintain close ties with their kin, and the social interactions between related animals may improve offspring survival. Poaching disrupts kin-related social groups with consequences for future population viability.

Overkill, or excessive human exploitation, will remain a problem for all animals and plants that are valuable or large. Large predators are particularly susceptible (Ripple et al. 2014b).

HABITAT DEGRADATION AND DESTRUCTION

The second factor that promotes extinctions is habitat loss. Habitats may simply be destroyed to make way for housing developments or agricultural fields. Cases of habitat destruction appear to provide the simplest examples of declining populations. The simplest model of habitat loss is that if you lose one-half of the habitat, you will have one-half of your original population. This is the "agricultural" model of habitat. If a farm is reduced in size by half, the farmer will typically be forced to reduce the number of cows he or she has to one-half their original number. But this agricultural model is completely wrong when applied to natural populations. Figure 12.1 illustrates the decline in bird species in the Brazilian Amazon when undisturbed primary forest is harvested, burnt, and converted to agriculture. As long as there is sufficient primary forest left in a landscape, forest birds will not go extinct, although their populations will be reduced. This raises the question of how much primary forest is enough to prevent extinctions, a difficult question for land planners.

The global problem is that humans have appropriated a large fraction

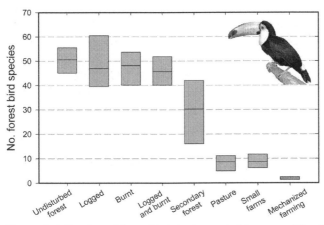

Figure 12.1 Number of forest birds in the central Amazon of Brazil in undisturbed forest and in areas disturbed by forestry and agriculture. Counts were made on 165 transects in the Santarém/Belterra district in central Pará state. The line in each rectangle shows the average number of bird species for each land use category and the grey box includes 50% of all the individual counts of birds for that land use. (Data from Moura et al. 2013.)

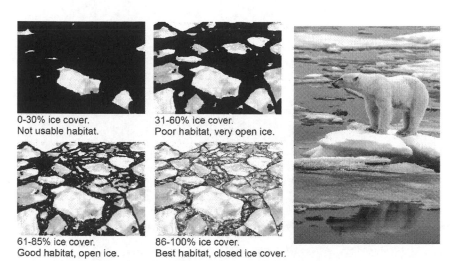

0-30% ice cover.
Not usable habitat.

31-60% ice cover.
Poor habitat, very open ice.

61-85% ice cover.
Good habitat, open ice.

86-100% ice cover.
Best habitat, closed ice cover.

Figure 12.2 Polar bear sea ice habitat classes in the eastern Canadian arctic based on the area of ocean covered by ice. Polar bears feed almost exclusively on seals and live most of their lives on sea ice. If there is not enough sea ice cover, seals move elsewhere and polar bears starve. Ice cover in polar seas can be monitored readily from satellites. (Modified from Sahanatien and Derocher 2012.)

of the land surface of the Earth for agriculture, and many plants and animals cannot survive in an agricultural landscape. Of the remaining areas, many have been fragmented, or broken up into small patches, a common situation in every country on Earth (Echeverria et al. 2006). Habitat fragmentation has many components with varying effects on populations. A particularly simple case is that of the endangered polar bear of the arctic. Suitable habitat for polar bears is sea ice, and this is because they feed almost entirely on seals that reproduce and live on sea ice. If the sea ice melts, seals move further north and polar bears must follow or starve. Extensive studies have shown that about 50–60% sea ice cover is required to support polar bears, and from satellite data it is relatively easy to classify habitat quality for polar bears (Figure 12.2). The result of these studies has been the recognition that the southern parts of the geographic range of polar bears are becoming uninhabitable because sea ice is too sparse even during winter (Stirling 2011). If global warming continues, and polar sea ice continues to shrink, the polar bear population will be under severe threat.

Fragmentation of habitats can result in species loss, and this can be seen by measuring the occupancy of species in small habitat patches.

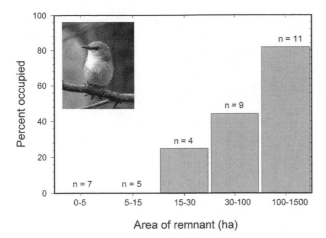

Figure 12.3 The percentage of remnant woodlands that were occupied by the eastern yellow robins (*Eopsaltria australis*) in central New South Wales, Australia. Small woodlots are not useful for the conservation of this small bird. (Data from Briggs et al. 1999, photo courtesy of Peter Fullagar.)

At one extreme, when patches are too small the species cannot survive. We can see this very clearly by looking at the occupancy rate of a species in habitats of differing size. Birds have been studied extensively for the effects of fragmentation. Figure 12.3 illustrates that for the eastern yellow robin in southeastern Australia, occupancy is maximal only in forest remnants that are more than 100 hectares. There is in general a good relationship between the body size of animals and the area required for survival and reproduction (Biedermann 2003). Larger animals need a larger area of habitat.

Recolonization may not always occur in isolated patches. A classic case is the Bogor Botanical Garden near Jakarta in Indonesia. The Bogor Botanical Garden was established in 1817 on 86 hectares in west Java. Until 1936 the botanical garden was connected with other forest areas to the east, but for the past 80 years it has been an isolated patch of forest with the nearest patch 5 kilometers away (Sodhi et al. 2006). Of the 97 bird species recorded as breeding in the botanical garden during 1932–1952, 57 species (59%) had disappeared by 2004. Many of the larger birds were lost, and their low abundance combined with the lack of recolonization from surrounding areas has been the main cause of extinction. The result is that much of the conservation value of the Botanical Garden for birds has been lost because it is too small by itself to support a secure population of many tropical forest birds.

In almost all cases habitat fragmentation leads to species loss. The prairies of North America are a good example. Prairie covered about 800,000 hectares of southern Wisconsin when Europeans first arrived, and now prairie occupies less than 0.1% of its original area (Leach and Givnish 1996). Plant surveys of 54 Wisconsin prairie remnants studied between 1948 and 1954 were repeated in 1987–88. Between 8% and 60% of the plant species were lost during these four decades, at average rates between 0.5% and 1.0% per year. At this rate of extinction approximately half the plant species would disappear in 50 to 100 years. Losses were particularly high among the shorter plant species and the rare species. The control of fire in prairies seems to be the agent of decline for prairie plants, and controlled burns should be done to reverse these population declines.

IMPACTS OF INTRODUCED PEST SPECIES

Introduced animals are responsible for about 40% of historic extinctions. Most of the collected data involve mammals and birds, for which we have more detailed information, and we do not know if this same percentage would apply to invertebrates and plants. But no one doubts the adverse effects of some introduced species. The Nile perch, which was introduced into Lake Victoria in the early 1980s, caused the extinction or near-extinction of 500 species of cichlid fish found only in that lake (Seehausen et al. 1997). The Nile perch was then subjected to a large fishery, and its abundance was greatly reduced during the 1990s and 2000s. This reduction in predation pressure has resulted in a few of the endemic cichlid fishes recovering in Lake Victoria, but most of the endemic fishes there have disappeared or become extremely rare (Witte et al. 2013).

Nearly 50% of the mammal extinctions of the past 200 years occurred in Australia. Neither very small nor very large mammals have been affected in these recent losses. All the missing mammals fall within a critical weight range from 35 to 4,200 grams (Burbidge and McKenzie 1989). Many causes can be suggested to explain these extinctions, from habitat clearing associated with agriculture, to changes in the fire regime, to introduced herbivores as competitors, to introduced predators. The main culprit seems to be introduced predators, particularly the red fox (Short et al. 2002). The details of the loss of medium-sized marsupials in Australia is a mirror image of the spread of the red fox. If the red fox can be controlled, some of the threatened species, now confined to offshore fox-free islands, could be reintroduced to their former range.

Introduced species are one of the most serious conservation problems today and the leading cause of animal extinctions (Szabo et al. 2012). For birds, it is estimated that 141 species have gone extinct around the globe since AD 1500. These extinctions were concentrated in Hawaii, Australia,

New Zealand, and Polynesia. Introduced species are the principal cause of 58% of these bird extinctions; and the percentages for fish and mammals are nearly the same. As global trade has increased during the last century, many inadvertent or deliberate introductions have been occurring, often with little regard for their conservation consequences.

CASCADING EXTINCTIONS

The last cause of extinctions is food web collapse in which primary extinctions lead to secondary extinctions; that is, one extinction can cause another. If other species depend on a lost species for survival, these other species must also go extinct. Chains of extinctions require specialist species that depend on only one other species for food or shelter, the kinds of food web relationships that are more typical of tropical areas than of temperate or polar zones. One obvious cascade of extinctions involves the loss of parasite species when their host goes extinct. This matter has received scant attention to date and there are few well-documented examples.

The clearest examples of chains of extinctions involve large predators that disappeared when their prey went extinct. The extinct forest eagle of New Zealand (*Harpagornis moorei*), which weighed 10–13 kilograms and preyed on large ground birds, died out around AD 1400, when moas became extinct in New Zealand (Holdaway 1989). The decline of the black-footed ferret in North America was associated with the decline of its main food, prairie dogs, on the Great Plains. Currently the black-footed ferret is being reintroduced into areas where prairie dog colonies are safe (Miller and Reading 2012), but its future is not secure because prairie dogs are often persecuted by ranchers in the western US. The black-footed ferret is also highly susceptible to canine distemper, a disease endemic in carnivores on the Great Plains of North America.

Without a clear understanding of the food web dynamics of an ecosystem, we cannot make predictions about what secondary extinctions will follow from the loss of any primary species. We have proceeded on the assumption that each species that is lost is an independent entity and its disappearance on a local or global scale will not greatly affect the existence of other species in the food web. This assumption is in need of much further study and at present we cannot make predictions about the consequences of the loss of many endangered species.

ROLE OF NATIONAL PARKS AND RESERVES

One way to conserve species that could be endangered is to set up national parks, reserves, or protected areas. National parks in many countries have been viewed as protected areas for populations and communities. If a national park or reserve is to be effective for conservation, we need to specify exactly what it is supposed to accomplish. Two quite divergent aims are often stated for parks and reserves:

1 To protect specific animal and plant communities subject to change because of fire, grazing, or predation. Management of these reserves requires the setting of the permissible levels of fire, grazing, and predation.

2 To allow the system to exist in its natural state and to change as governed by undisturbed ecological processes, so that no attempt will be made to influence the resulting changes in populations and communities.

Often reserves like national parks have both these aims, and a recipe for conflict exists, often over what kinds of changes are acceptable and what uses are unacceptable to the reserve managers or the general public.

Few national parks and reserves are set up to exist in their natural state without any management. Many national parks around the Earth on land are already established, and it seems unlikely that much more land will be newly designated as a national park. In the marine realm, marine protected areas are now being established, and more planning needs to be done in their selection. Two key questions about marine reserves are where to locate them (if there is a choice) and what size is needed to achieve conservation goals.

About 14.6% of the Earth's land area, 9.7% of coastal waters, and 2.3% of the open ocean area were protected as of 2010 (Bertzky et al. 2012). The nationally designated protected terrestrial areas covered 17 million square kilometers, an area twice the size of Brazil. Marine protection is still concentrated in the near-coastal areas (0–12 nautical miles, or 0–22 kilometers, from land). The goal of many governments is to protect about 17% of terrestrial habitats and to increase the area of coastal waters that are protected. If we are given the job of adding to existing reserves, how should we proceed? One way is to identify hotspots that are particularly rich in species, and to locate reserves in these areas. One problem with this approach is that areas that are hotspots for birds are typically not hotspots for butterflies, so that one cannot choose reserves on the basis of only one

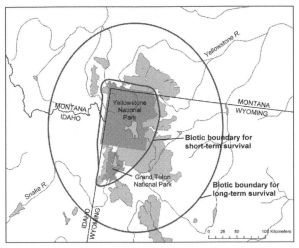

Figure 12.4 The legal and conservation-defined boundaries of the Yellowstone–Grand Teton National Park assemblage in the United States with respect to grizzly bears (*Ursus arctos*). The conservation-defined boundaries (thick lines) are defined as the area necessary to support a population of 50 bears for short-term survival or 500 individuals for long-term survival (*Ursus arctos*), which has a large home range (489 km²). National parks in dark grey, federal wilderness areas in light grey. (Modified from Newmark 1985.)

taxonomic group and hope that it will protect other groups as well. Nevertheless, some small areas are much richer in species than others, and we should use this kind of information to help select additional reserves. For a national park or reserve system to be useful for conservation, it is necessary to know the ecological requirements of the species of concern. A special problem exists for species that use temporary habitats. Many butterflies use areas for egg laying and larval development that are temporary. If the habitat is set aside in a reserve and the protected area changes, for example from a meadow to a forest, the butterfly would lose its host plants and disappear. Butterflies and birds in particular often need several kinds of habitats, and movements between suitable patches of habitat are critical to survival.

One of the most significant contributions of conservation biology has been to show that the minimum viable populations of some species are large so that it may be impossible to maintain them in national parks or sanctuaries without considering the land use just outside the park. Figure 12.4 illustrates this problem for the grizzly bear in the Yellowstone–Grand Teton Park area of the western USA. The Yellowstone Park region is among the largest of the parks in the United States, and everyone has always assumed that it was sufficiently large to preserve all of its biodiversity. Not so. The biotic boundary of the area needed to support a minimum viable population of 500 grizzly bears encloses 122,330 square kilometers, about 12 times the actual park area of 10,328 square kilometers. Our existing parks are far too small to maintain large mammals and birds. Areas of private land outside of parks must also contribute to the preservation of biodiversity. The important message is that for the conservation of the Earth's biodiversity, we cannot rely only on parks and reserves. We need to cultivate methods of preserving biodiversity in all areas of human use—agricultural regions, grazing areas, and forest plantations.

GENETIC PROBLEMS IN THREATENED SPECIES

Most problems with declining and threatened species are considered to be ecological, and the solution lies in finding the threatening processes for each particular species and neutralizing these by management actions. But there are potential genetic problems involved with populations that are small and declining, and genetics can thus contribute to extinctions. The processes involved are summarized in the extinction vortex (Figure 12.5). This approach to extinction through genetics focuses on the population consequences of having a small population. Examples of where this approach could be useful for conservation might be a small island population of a bird, or a small group of endangered species of animals like a monkey that exists now only in a zoo or a rare orchid in a botanical garden.

Small populations risk positive feedback loops of inbreeding depression, genetic drift, and demographic chance events that lead inexorably to extinction. The key element is the maintenance of genetic variability, and the critical assumption is that species require genetic variability for future evolution and thus long-term persistence. The flip side of this assumption is that species that have no genetic variability will never persist in evolutionary time. If this assumption is correct, conservation biologists must strive above all to maintain genetic variability in threatened populations (Johnson and Dunn 2006).

What factors might cause a species with a small population to go extinct? Once a population is small, it is subject to a variety of chance events. Two kinds of chance events can contribute to extinction.

1   *Demographic variability.* Random variation in birth and death rates can lead by chance to extinction. If only a few individuals

make up the population, the fate of each individual can be critical to population survival. Consider the extreme case of an island population with only one male and one female. If the female produces only male offspring and then dies, this hypothetical population goes extinct. In general demographic variability is critical to extinction only when populations are less than about 30–50 individuals.

2 *Genetic variability*. Since evolution cannot occur without genetic variability, any loss of genetic variation might cause extinction. Many genetic studies have shown that individuals with more genetic variability are more fit than individuals with less genetic variability. Genetic variability can be lost by chance during bisexual reproduction (since only one-half of one's genes are passed on) and by inbreeding (in which the two parents are close relatives). Both sources of loss are minimized when breeding populations become large; these are the classic examples of problems that can affect only small populations.

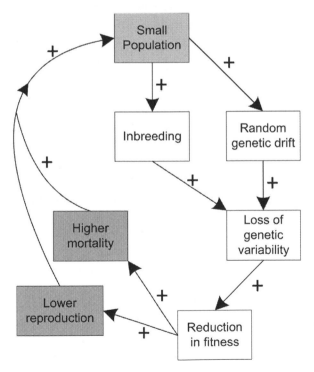

Figure 12.5 The extinction vortex for small-threatened populations. Small populations such as those on islands or in zoos, fall into a vortex of positive feedback loops in which small population size leads to inbreeding and genetic drift, and the loss of genetic variability. Since genetic variability is necessary for viability, fitness falls and the population size is reduced further because mortality goes up and reproduction goes down. The final result can be extinction unless new genetic variation can be brought into the population.

Small populations could also go extinct because of natural catastrophes like volcanic eruptions, floods, fires, and hurricanes. Species that have few individuals in only a very few localities are always highly susceptible to chance catastrophes.

All of these threatening processes must be alleviated in reversing extinction threats. A good example of current concern is the Spanish imperial eagle (*Aquila adalberti*), one of the most endangered birds of prey in the world. The population of the eagles in Doñana National Park (southwestern Spain) has suffered a dramatic decline from 1992 to 2003 after a long period of stability from the 1970s. High adult mortality caused by poisoning from agricultural insecticides was the main cause of the decline. The declining population entered the extinction vortex with decreasing fecundity and an offspring sex ratio biased to males (Ferrer at al. 2013). In face of the imminent extinction, an urgent conservation plan was implemented in 2004 (Figure 12.6). Supplementary feeding throughout the year with live wild rabbits was undertaken to prevent breeders from foraging outside the national park, thereby reducing adult mortality from poisoning. The population was reinforced with the release of 15 young eagles (mainly captive-bred females from other areas). After implementing the plan, the annual adult mortality decreased, increasing the average age of breeders. Accordingly, the fecundity recovered to values close to those prior to decline (from 0.6 to 1.3 young/pair), and the sex ratio was balanced again. These conservation actions were effective in a short time and made possible the rescue of the threatened population from the extinction vortex. The need now is to reduce pesticide use in agriculture to stop the accidental poisoning of eagles.

CONCLUSIONS

Conservation biology is focused on the ecology of rare and declining species. The declining population approach focuses on identifying the ecological causes of decline and designing alleviation measures to stop the decline. It contains almost no ecological theory but is focused on individual action plans. Only by understanding the population biology of an endangered plant or animal can we provide a rescue plan for a declining population. In some cases, like the African elephant, the causes of population decline are clear. In other cases we do not have the ecological understanding to recommend action, and we need to develop insights for action plans.

Extinction is the ultimate focus of conservation. The major causes of re-

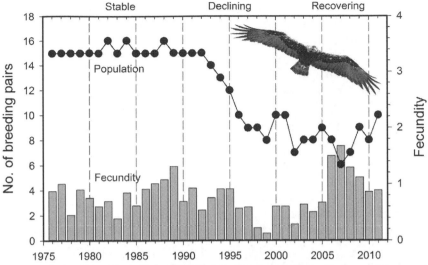

Figure 12.6 Number of breeding pairs (black line) and mean fecundity (grey histograms) in the population of Spanish imperial eagle in Doñana National Park (south-western Spain). During the period of population stability (1976–1991) the population remained close to the carrying capacity of the park (16 pairs) with stable mean fecundity of 1 offspring per pair per year. In the declining period (1992–2003), the number of breeding pairs dropped off as a result of increasing adult mortality by poisoning, which in turn led to the reduction of fecundity because older birds are more successful than younger birds. In 2004, a multi-action recovery program was applied. Fecundity showed an immediate and noticeable increase followed by a delayed population recovery. (Data from Ferrer et al. 2013.)

cent extinctions in all plant and animal groups are habitat destruction and introduced species. Habitat destruction leads to population reductions that may trigger the extinction vortex, so protecting habitat is a major goal for all conservation efforts.

Existing parks and reserves are seldom large enough to contain viable populations of larger vertebrates, and conservation efforts on private lands are essential to maintaining our flora and fauna.

The ecological challenge of today is to develop specific action plans for individual threatened species, and to act on the processes that threaten extinction. Without parks and reserves there can be no conservation, but they are not sufficient; there is no guarantee of success unless we can solve the challenging problems of endangered species.

REFERENCES

Abella, S. R. (2010) Disturbance and plant succession in the Mojave and Sonoran deserts of the American Southwest. *International Journal of Environmental Research and Public Health*, 7, 1248–1284.

Abraham, K. F., Jefferies, R. L. & Alisauskas, R. T. (2005) The dynamics of landscape change and snow geese in mid-continent North America. *Global Change Biology*, 11, 841–855.

Ahlgren, I., Frisk, T. & Kamp-Nielsen, L. (1988) Empirical and theoretical models of phosphorus loading, retention and concentration vs. lake trophic status. *Hydrobiologia*, 170, 285–303.

Alisauskas, R. T., Rockwell, R. F., Dufour, K. W., Cooch, E. G., Zimmerman, G., Drake, K. L., Leafloor, J. O., Moser, T. J. & Reed, E. T. (2011) Harvest, survival, and abundance of midcontinent Lesser Snow Geese relative to population reduction efforts. *Wildlife Monographs*, 179, 1–42.

Allen, K. R. (1980) *Conservation and Management of Whales*. University of Washington Press, Seattle.

Allendorf, F. W. & Hard, J. J. (2009) Human-induced evolution caused by unnatural selection through harvest of wild animals. *Proceedings of the National Academy of Sciences of the USA*, 106, 9987–9994.

Alongi, D. M. (2008) Mangrove forests: resilience, protection from tsunamis, and responses to global climate change. *Estuarine, Coastal and Shelf Science*, 76, 1–13.

Amthor, J. S. & Baldocchi, D. D. (2001) Terrestrial higher plant respiration and net primary production. *Terrestrial Global Productivity* (eds. J. Roy, B. Saugier & H. A. Mooney), pp. 33–59. Academic Press, San Diego.

Andersson, D. I. (2006) The biological cost of mutational antibiotic resistance: any practical conclusions? *Current Opinion in Microbiology* 9, 461–465.

Archie, E. A. & Chiyo, P. I. (2012) Elephant behaviour and conservation: social relationships, the effects of poaching, and genetic tools for management. *Molecular Ecology*, 21, 765–778.

Barnosky, A. D., Koch, P. L., Feranec, R. S., Wing, S. L. & Shabel, A. B. (2004) Assessing the causes of Late Pleistocene extinctions on the continents. *Science*, 306, 70–75.

Barnosky, A. D. & Lindsey, E. L. (2010) Timing of Quaternary megafaunal extinction in South America in relation to human arrival and climate change. *Quaternary International*, 217, 10–29.

Beckie, H. J., Lozinski, C., Shirriff, S. & Brenzil, C. A. (2013) Herbicide-resistant weeds in the Canadian prairies: 2007 to 2011. *Weed Technology*, 27, 171–183.

Bell, G. (2013) Evolutionary rescue and the limits of adaptation. *Philosophical Transactions of the Royal Society B: Biological Sciences*, 368, 20120080.

Benson, L. V., Pauketat, T. R. & Cook, E. R. (2009) Cahokia's boom and bust in the context of climate change. *American Antiquity*, 74, 467–483.

Bertzky, B., Corrigan, C., Kemsey, J., Kenney, S., Ravilious, C., Besançon, C. & Burgess, N. (2012) Protected Planet Report 2012: Tracking progress towards global targets for protected areas. International Union for the Conservation of Nature and United Nations Environment Program, IUCN, Gland, Switzerland, and UNEP-WCMC, Cambridge, UK.

Bianchi, T. S., DiMarco, S. F., Cowan, J. H., Jr., Hetland, R. D., Chapman, P., Day, J. W. & Allison, M. A. (2010) The science of hypoxia in the Northern Gulf of Mexico: a review. *Science of the Total Environment*, 408, 1471–1484.

Biedermann, R. (2003) Body size and area-incidence relationships: is there a general pattern? *Global Ecology & Biogeography*, 12, 381–387.

Bjorkman, A.D. & Vellend, M. (2010) Defining historical baselines for conservation: ecological changes since European settlement on Vancouver Island, Canada. *Conservation Biology*, 24, 1559–1568.

Björkman, O. & Berry, J. (1973) High efficiency photosynthesis. *Scientific American*, 229, 80–93.

Blackburn, T.M., Prowse, T.A.A., Lockwood, J.L. & Cassey, P. (2013) Propagule pressure as a driver of establishment success in deliberately introduced exotic species: fact or artefact? *Biological Invasions*, 15, 1459–1469.

Blake, S., Strindberg, S., Boudjan, P., Makombo, C., Bila-Isia, I., Ilambu, O., Grossmann, F., Bene-Bene, L., de Semboli, B., Mbenzo, V., S'Hwa, D., Bayogo, R., Williamson, L., Fay, M., Hart, J. & Maisels, F. (2007) Forest elephant crisis in the Congo Basin. *PLoS Biology*, 5, 0945–0953.

Blanc, J. J., Barnes, R. F. W., Craig, G. C., Dublin, H. T., Thouless, C. R., Douglas-Hamilton, I. & Hart, J. A. (2007) African Elephant Status Report 2007: an update from the African Elephant Database. *Occasional Paper Series of the IUCN Species Survival Commission* 33, 1–276.

Blankenship, R. E. (2002) *Molecular Mechanisms of Photosynthesis*. Blackwell Science, Oxford, U.K.

Blankenship, R. E., Tiede, D. M., Barber, J., Brudvig, G. W., Fleming, G., Ghirardi, M., Gunner, M. R., Junge, W., Kramer, D. M., Melis, A., Moore, T. A., Moser, C. C., Nocera, D. G., Nozik, A. J., Ort, D. R., Parson, W. W., Prince, R. C. & Sayre, R. T. (2011) Comparing photosynthetic and photovoltaic efficiencies and recognizing the potential for improvement. *Science*, 332, 805–809.

Bourguet, D., Delmotte, F., Franck, P. & Consortium, R. (2013) Heterogeneity of selection and the evolution of resistance. *Trends in Ecology & Evolution*, 28, 110–118.

Boyd, P. W., Jickells, T., Law, C. S., Blain, S., Boyle, E. A., Buesseler, K. O., Coale, K. H., Cullen, J. J., de Baar, H. J. W., Follows, M., Harvey, M., Lancelot, C.,

Levasseur, M., Owens, N. P. J., Pollard, R., Rivkin, R. B., Sarmiento, J., Schoemann, V., Smetacek, V., Takeda, S., Tsuda, A., Turner, S. & Watson, A. J. (2007) Mesoscale iron enrichment experiments 1993–2005: synthesis and future directions. *Science*, 315, 612–617.

Briggs, S.V., Seddon, J. & Doyle, S. (1999) Predicting biodiversity of woodland remnants for on-ground conservation. *National Heritage Trust (Australia), Special Report AA 1373.97, Canberra, September 1999*.

Bristow, C. S., Hudson-Edwards, K. A. & Chappell, A. (2010) Fertilizing the Amazon and equatorial Atlantic with West African dust. *Geophysical Research Letters*, 37, L14807.

Brower, L. P., Taylor, O. R., Williams, E. H., Slayback, D. A., Zubieta, R. R. & Ramírez, M. I. (2012) Decline of monarch butterflies overwintering in Mexico: is the migratory phenomenon at risk? *Insect Conservation & Diversity*, 5, 95–100.

Buckle, A. (2013) Anticoagulant resistance in the United Kingdom and a new guideline for the management of resistant infestations of Norway rats (*Rattus norvegicus* Berk.). *Pest Management Science*, 69, 334–341.

Buesseler, K. O., Andrews, J. E., Pike, S. M. & Charette, M. A. (2004) The effects of iron fertilization on carbon sequestration in the Southern Ocean. *Science*, 304, 414–417.

Burbidge, A. A. & McKenzie, N. L. (1989) Patterns in the modern decline of Western Australia's vertebrate fauna: causes and conservation implications. *Biological Conservation*, 50, 143–198.

Burrows, M. T., Schoeman, D. S., Richardson, A. J., Molinos, J. G., Hoffmann, A., Buckley, L. B., Moore, P. J., Brown, C. J., Bruno, J. F., Duarte, C. M., Halpern, B. S., Hoegh-Guldberg, O., Kappel, C. V., Kiessling, W., O'Connor, M. I., Pandolfi, J. M., Parmesan, C., Sydeman, W. J., Ferrier, S., Williams, K. J. & Poloczanska, E. S. (2014) Geographical limits to species-range shifts are suggested by climate velocity. *Nature*, 507, 492–495.

Cahill, A. E., Aiello-Lammens, M. E., Fisher-Reid, M. C., Hua, X., Karanewsky, C. J., Ryu, H. Y., Sbeglia, G. C., Spagnolo, F., Waldron, J. B. & Wiens, J. J. (2014) Causes of warm-edge range limits: systematic review, proximate factors and implications for climate change. *Journal of Biogeography*, 41, 429–442.

Cargill, S. M. & Jefferies, R. L. (1984) The effects of grazing by lesser snow geese on the vegetation of a sub-arctic salt marsh. *Journal of Applied Ecology*, 21, 669–686.

Cerrato, M. E. & Blackmer, A. M. (1990) Comparison of models for describing corn yield response to nitrogen fertilizer. *Agronomy Journal*, 82, 138–143.

Chauhan, B. S. (2012) Weed ecology and weed management strategies for dry-seeded rice in Asia. *Weed Technology*, 26, 1–13.

Chauhan, B. S. (2013) Strategies to manage weedy rice in Asia. *Crop Protection*, 48, 51–56.

Chen, I.-C., Hill, J. K., Ohlemüller, R., Roy, D. B. & Thomas, C. D. (2011) Rapid range shifts of species associated with high levels of climate warming. *Science*, 333, 1024–1026.

Clark, G. F., Stark, J. S., Johnston, E. L., Runcie, J. W., Goldsworthy, P. M., Raymond, B. & Riddle, M. J. (2013) Light-driven tipping points in polar ecosystems. *Global Change Biology*, 19, 3749–3761.

Comte, L. & Grenouillet, G. (2013) Do stream fish track climate change? Assessing distribution shifts in recent decades. *Ecography*, 36, 1236–1246.

Connell, J. H., Hughes, T. P. & Wallace, C. C. (1997) A 30-year study of coral abundance, recruitment, and disturbance at several scales in space and time. *Ecological Monographs*, 67, 461–488.

Costanza, J. K., Weiss, J. & Moody, A. (2013) Examining the knowing-doing gap in the conservation of a fire-dependent ecosystem. *Biological Conservation*, 158, 107–115.

Davies, J. & Davies, D. (2010) Origins and evolution of antibiotic resistance. *Microbiology and Molecular Biology Review*, 74, 417–433.

De Rose, R. C., Oguchi, T., Morishima, W. & Collado, M. (2011) Land cover change on Mt. Pinatubo, the Philippines, monitored using ASTER VNIR. *International Journal of Remote Sensing*, 32, 9279–9305.

DeAngelis, D. L. (1992) *Dynamics of Nutrient Cycling and Food Webs*. Chapman & Hall, New York.

Del Moral, R., Thomason, L. A., Wenke, A. C., Lozanof, N. & Abata, M. D. (2012) Primary succession trajectories on pumice at Mount St. Helens, Washington. *Journal of Vegetation Science*, 23, 73–85.

Del Moral, R. & Wood, D. M. (1993) Early primary succession on the volcano Mount St. Helens. *Journal of Vegetation Science*, 4, 223–234.

Diamond, J. (2005) *Collapse: How Societies Choose to Fail or Succeed*. Viking, New York.

Doering, P. H., Oviatt, C. A., Nowicki, B. L., Klos, E. G. & Reed, L. W. (1995) Phosphorus and nitrogen limitation of primary production in a simulated estuarine gradient. *Marine Ecology Progress Series*, 124, 271–287.

Donelson, J. M., Munday, P. L., McCormick, M. I. & Pitcher, C. R. (2012) Rapid transgenerational acclimation of a tropical reef fish to climate change. *Nature Climate Change*, 2, 30–32.

Downing, J. A., Osenberg, C. W. & Sarnelle, O. (1999) Meta-analysis of marine nutrient-enrichment experiments: variation in the magnitude of nutrient limitation. *Ecology*, 80, 1157–1167.

Dublin, H. T. (2013) African Elephant Specialist Group report. *Pachyderm*, 54, 1–7.

Ebert, D. & Bull, J. J. (2003) Challenging the trade-off model for the evolution of virulence: is virulence management feasible? *Trends in Microbiology*, 11, 15–20.

Echeverria, C., Coomes, D., Salas, J., Rey-Benayas, J. M., Lara, A. & Newton, A.

(2006) Rapid deforestation and fragmentation of Chilean temperate forests. *Biological Conservation*, 130, 481–494.

Edgar, G. J., Stuart-Smith, R. D., Willis, T. J., Kininmonth, S., Baker, S. C., Banks, S., Barrett, N. S., Becerro, M. A., Bernard, A. T. F., Berkhout, J., Buxton, C. D., Campbell, S. J., Cooper, A. T., Davey, M., Edgar, S. C., Forsterra, G., Galvan, D. E., Irigoyen, A. J., Kushner, D. J., Moura, R., Parnell, P. E., Shears, N. T., Soler, G., Strain, E. M. A. & Thomson, R. J. (2014) Global conservation outcomes depend on marine protected areas with five key features. *Nature*, 506, 216–220.

Edmondson, W. T. (1991) *The Uses of Ecology*. University of Washington Press, Seattle.

Eisenberg, C., Seager, S. T. & Hibbs, D. E. (2013) Wolf, elk, and aspen food web relationships: context and complexity. *Forest Ecology and Management*, 299, 70–80.

Estes, J. A., Tinker, M. T., Williams, T. M. & Doak, D. F. (1998) Killer whale predation on sea otters linking oceanic and nearshore ecosystems. *Science*, 282, 473–476.

Fageria, N. K., Moraes, M. F., Ferreira, E. P. B. & Knupp, A. M. (2012) Biofortification of trace elements in food crops for human health. *Communications in Soil Science and Plant Analysis*, 43, 556–570.

Fenner, F. & Ratcliffe, F. N. (1965) *Myxomatosis*. Cambridge University Press, Cambridge, U.K.

Ferrer, M., Newton, I. & Muriel, R. (2013) Rescue of a small declining population of Spanish imperial eagles. *Biological Conservation*, 159, 32–36.

Fox, H. E. & Caldwell, R. L. (2006) Recovery from blast fishing on coral reefs: a tale of two scales. *Ecological Applications*, 16, 1631–1635.

Fritts, H. C. (1976) *Tree Rings and Climate*. Academic Press, London.

Fryxell, J. M., Greever, J. & Sinclair, A. R. E. (1988) Why are migratory ungulates so abundant? *American Naturalist*, 131, 781–798.

Good, N. F. (1968) A study of natural replacement of chestnut in six stands in the highlands of New Jersey. *Bulletin of the Torrey Botanical Club*, 95, 240–253.

Goward, S. N., Tucker, C. J. & Dye, D. G. (1985) North American vegetation patterns observed with the NOAA-7 advanced very high resolution radiometer. *Vegetatio*, 64, 3–14.

Graham, N. A. J., Nash, K. L. & Kool, J. T. (2011) Coral reef recovery dynamics in a changing world. *Coral Reefs*, 30, 283–294.

Grant, S. M., Hill, S. L., Trathan, P. N. & Murphy, E. J. (2013) Ecosystem services of the Southern Ocean: trade-offs in decision-making. *Antarctic Science*, 25, 603–617.

Grudd, H. (2008) Torneträsk tree-ring width and density AD 500–2004: a test of climatic sensitivity and a new 1500-year reconstruction of north Fennoscandian summers. *Climate Dynamics*, 31, 843–857.

Gunn, J. M. & Mills, K. H. (1998) The potential for restoration of acid-damaged lake trout lakes. *Restoration Ecology*, 6, 390–397.

Hare, S. R. & Mantua, N. J. (2000) Empirical evidence for North Pacific regime shifts in 1977 and 1989. *Progress in Oceanography*, 47, 103–145.

Haydon, D. T., Shaw, D. J., Carradori, I. M., Hudson, P. J. & Thirgood, S. J. (2002) Analysing noisy time-series: describing regional variation in the cyclic dynamics of red grouse. *Proceedings of the Royal Society of London, Series B*, 269, 1609–1617.

Heap, I. & LeBaron, H. (2001) Introduction and overview of resistance. *Herbicide Resistance and World Grains* (eds. S. B. Powles & D. L. Shaner), pp. 1–22. CRC Press, Boca Raton, FL.

Helander, M., Saloniemi, I. & Saikkonen, K. (2012) Glyphosate in northern ecosystems. *Trends in Plant Science*, 17, 569–574.

Herweijer, C., Seager, R., Cook, E. R. & Emile-Geay, J. (2007) North American droughts of the last millennium from a gridded network of tree-ring data. *Journal of Climate*, 20, 1353–1376.

Hilborn, R. (2007) Moving to sustainability by learning from successful fisheries. *Ambio*, 36, 1–9.

Hobbs, R. J. & Cramer, V. A. (2008) Restoration ecology: interventionist approaches for restoring and maintaining ecosystem function in the face of rapid environmental change. *Annual Review of Environment and Resources*, 33, 39–61.

Holdaway, R. N. (1989) New Zealand's pre-human avifauna and its vulnerability. *New Zealand Journal of Ecology*, 12, 11–25.

Holt, B. G., Lessard, J.-P., Borregaard, M. K., Fritz, S. A., Araújo, M. B., Dimitrov, D., Fabre, P.-H., Graham, C. H., Graves, G. R., Jønsson, K. A., Nogués-Bravo, D., Wang, Z., Whittaker, R. J., Fjeldså, J. & Rahbek, C. (2013) An update of Wallace's Zoogeographic Regions of the World. *Science*, 339, 74–78.

Hong, Y. T., Hong, B., Lin, Q. H., Zhu, Y. X., Shibata, Y., Hirota, M., Uchida, M., Leng, X. T., Jiang, H. B., Xu, H., Wang, H. & Yi, L. (2003) Correlation between Indian Ocean summer monsoon and North Atlantic climate during the Holocene. *Earth and Planetary Science Letters*, 211, 371–380.

Horne, A. J. & Goldman, C. R. (1994) *Limnology*. McGraw-Hill, New York.

Hughes, T. P., Graham, N. A. J., Jackson, J. B. C., Mumby, P. J. & Steneck, R. S. (2010) Rising to the challenge of sustaining coral reef resilience. *Trends in Ecology & Evolution*, 25, 633–642.

Hutchings, M. J. (2010) The population biology of the early spider orchid *Ophrys sphegodes* Mill. III: demography over three decades. *Journal of Ecology*, 98, 867–878.

Intergovernmental Panel on Climate Change (2013) IPCC Fifth Assessment Report: Climate Change 2013: The Physical Science Basis. Contribution of

Working Group I to the Fifth Assessment Report of the Intergovernmental Panel on Climate Change. Cambridge University Press, Cambridge, U.K.

Johnson, J. & Dunn, P. (2006) Low genetic variation in the heath hen prior to extinction and implications for the conservation of prairie-chicken populations. *Conservation Genetics*, 7, 37–48.

Johnson, S. R. & Cowan, I. M. (1974) Thermal adaptation as a factor affecting colonizing success of introduced Sturnidae (Aves) in North America. *Canadian Journal of Zoology*, 52, 1559–1576.

Leach, M. K. & Givnish, T. J. (1996) Ecological determinants of species loss in remnant prairies. *Science*, 273, 1555–1558.

Legagneux, P., Gauthier, G., Berteaux, D., Bêty, J., Cadieux, M.-C., Bilodeau, F., Bolduc, E., McKinnon, L., Tarroux, A., Therrien, J.-F., Morissette, L. & Krebs, C. J. (2012) Disentangling trophic relationships in a high arctic tundra ecosystem through food web modeling. *Ecology*, 93, 1707–1716.

Legge, S., Kennedy, M. S., Lloyd, R. A. Y., Murphy, S. A. & Fisher, A. (2011) Rapid recovery of mammal fauna in the central Kimberley, northern Australia, following the removal of introduced herbivores. *Austral Ecology*, 36, 791–799.

Likens, G. E., Bormann, F. H., Johnson, N. M., Fisher, D. W. & Pierce, R. S. (1970) Effects of forest cutting and herbicide treatment on nutrient budgets in the Hubbard Brook watershed-ecosystem. *Ecological Monographs*, 40, 23–47.

Link, J. (2002) Does food web theory work for marine ecosystems? *Marine Ecology Progress Series*, 230, 1–9.

Lipsett, J. & Simpson, J. (1973) Analysis of the response by wheat to application of molybdenum in relation to nitrogen status. *Australian Journal of Experimental Agriculture*, 13, 563–566.

Lucas, J. D. & Lacourse, T. (2013) Holocene vegetation history and fire regimes of *Pseudotsuga menziesii* forests in the Gulf Islands National Park Reserve, southwestern British Columbia, Canada. *Quaternary Research*, 79, 366–376.

Ludwig, D., Hilborn, R. & Walters, C. (1993) Uncertainty, resource exploitation, and conservation: lessons from history. *Science*, 260, 17, 36.

MacDonald, W. L. (2003) Dominating North American forest pathology issues of the 20th century. *Phytopathology*, 93, 1039–1040.

Mathews, T. J. & MacDorman, M. F. (2012) Infant mortality statistics from the 2008 period: linked birth/infant death data set. *National Vital Statistics Reports*, 60, 1–28.

McCauley, D. J., Keesing, F., Young, T. P., Allan, B. F. & Pringle, R. M. (2006) Indirect effects of large herbivores on snakes in an African savanna. *Ecology*, 87, 2657–2663.

McCune, J. L., Pellatt, M. G. & Vellend, M. (2013) Multidisciplinary synthesis of long-term human-ecosystem interactions: a perspective from the Garry oak ecosystem of British Columbia. *Biological Conservation*, 166, 293–300.

Mennerat, A., Nilsen, F., Ebert, D. & Skorping, A. (2010) Intensive farming:

evolutionary implications for parasites and pathogens. *Evolutionary Biology*, 37, 59–67.

Middleton, G. (2012) Nothing lasts forever: environmental discourses on the collapse of past societies. *Journal of Archaeological Research*, 20, 257–307.

Miller, B. & Reading, R. P. (2012) Challenges to black-footed ferret recovery: protecting prairie dogs. *Western North American Naturalist*, 72, 228–240.

Monteith, K. L., Long, R. A., Bleich, V. C., Heffelfinger, J. R., Krausman, P. R. & Bowyer, R. T. (2013) Effects of harvest, culture, and climate on trends in size of horn-like structures in trophy ungulates. *Wildlife Monographs*, 183, 1–28.

Montevecchi, W. A. & Kirk, D. A. (1996) Great auk: *Pinguinus impennis*. *Birds of North America*, 260, 1–20.

Moura, N. G., Lees, A. C., Andretti, C. B., Davis, B. J. W., Solar, R. R. C., Aleixo, A., Barlow, J., Ferreira, J. & Gardner, T. A. (2013) Avian biodiversity in multiple-use landscapes of the Brazilian Amazon. *Biological Conservation*, 167, 339–348.

Myers, J. H. & Cory, J. S. (2013) Population cycles in forest Lepidoptera revisited. *Annual Review of Ecology, Evolution, and Systematics*, 44, 565–592.

Myers, J. H., Jackson, C., Quinn, H., White, S. R. & Cory, J. S. (2009) Successful biological control of diffuse knapweed, *Centaurea diffusa*, in British Columbia, Canada. *Biological Control*, 50, 66–72.

Neve, P., Vila-Aiub, M. & Roux, F. (2009) Evolutionary-thinking in agricultural weed management. *New Phytologist*, 184, 783–793.

Newmark, W. D. (1985) Legal and biotic boundaries of western North American national parks: a problem of congruence. *Biological Conservation*, 33, 197–208.

Normile, D. (2008) Driven to extinction. *Science*, 319, 1606–1609.

Orians, G. H. & Milewski, A. V. (2007) Ecology of Australia: the effects of nutrient-poor soils and intense fires. *Biological Reviews*, 82, 393–423.

Pace, M. L., Carpenter, S. R., Johnson, R. A. & Kurtzweil, J. T. (2013) Zooplankton provide early warnings of a regime shift in a whole lake manipulation *Limnology and Oceanography*, 58, 525–532.

Paine, R. T. (1966) Food web complexity and species diversity. *American Naturalist*, 100, 65–75.

Paine, R. T. (1974) Intertidal community structure: experimental studies on the relationship between a dominant competitor and its principal predator. *Oecologia*, 15, 93–120.

Paine, R. T., Tegner, M. J. & Johnson, E. A. (1998) Compounded perturbations yield ecological surprises. *Ecosystems*, 1, 535–545.

Parr, C. L., Gray, E. F. & Bond, W. J. (2012) Cascading biodiversity and functional consequences of a global change–induced biome switch. *Diversity and Distributions*, 18, 493–503.

Pauly, D., Hilborn, R. & Branch, T. A. (2013) Fisheries: does catch reflect abundance? *Nature*, 494, 303–306.

Peacock, S. J., Krkošek, M., Proboszcz, S., Orr, C. & Lewis, M. A. (2012) Cessation of a salmon decline with control of parasites. *Ecological Applications*, 23, 606–620.

Pearcy, R. W. & Ehleringer, J. (1984) Comparative ecophysiology of C_3 and C_4 plants. *Plant Cell and Environment*, 7, 1–13.

Pelz, H.-J., Rost, S., Hünerberg, M., Fregin, A., Heiberg, A.-C., Baert, K., MacNicoll, A. D., Prescott, C. V., Walker, A.-S., Oldenburg, J. & Müller, C. R. (2005) The genetic basis of resistance to anticoagulants in rodents. *Genetics*, 170, 1839–1847.

Pelz, H.-J., Rost, S., Müller, E., Esther, A., Ulrich, R. G. & Müller, C. R. (2012) Distribution and frequency of VKORC1 sequence variants conferring resistance to anticoagulants in *Mus musculus*. *Pest Management Science*, 68, 254–259.

Petit, J. R., Jouzel, J., Raynaud, D., Barkov, N. I., Barnola, J. M., Basile, I., Bender, M., Chappellaz, J., Davis, J., Delaygue, G., Delmotte, M., Kotlyakov, V. M., Legrand, M., Lipenkov, V., Lorius, C., Pépin, L., Ritz, C., Saltzman, E. & Stievenard, M. (1999) Climate and atmospheric history of the past 420,000 years from the Vostok Ice Core, Antarctica. *Nature*, 399, 429–436.

Piertney, S. B., Lambin, X., Maccoll, A. D. C., Lock, K., Bacon, P. J., Dallas, J. F., Leckie, F., Mougeot, F., Racey, P. A., Redpath, S. & Moss, R. (2008) Temporal changes in kin structure through a population cycle in a territorial bird, the red grouse *Lagopus lagopus scoticus*. *Molecular Ecology*, 17, 2544–2551.

Pleasants, J. M. & Oberhauser, K. S. (2013) Milkweed loss in agricultural fields because of herbicide use: effect on the monarch butterfly population. *Insect Conservation and Diversity*, 6, 135–144.

Population Reference Bureau (2013) World Population Data Sheet 2013. http://www.prb.org/Publications/Datasheets/2013/2013-world-population-data-sheet/data-sheet.aspx.

Rashid, A., Rafique, E., Bhatti, A. U., Ryan, J., Bughio, N. & Yau, S. K. (2011) Boron deficiency in rainfed wheat in Pakistan: incidence, spatial variability and management strategies. *Journal of Plant Nutrition*, 34, 600–613.

Rawat, N., Neelam, K., Tiwari, V. K. & Dhaliwal, H. S. (2013) Biofortification of cereals to overcome hidden hunger. *Plant Breeding*, 132, 437–445.

Reeves, J. M., Bostock, H. C., Ayliffe, L. K., Barrows, T. T., De Deckker, P., Devriendt, L. S., Dunbar, G. B., Drysdale, R. N., Fitzsimmons, K. E., Gagan, M. K., Griffiths, M. L., Haberle, S. G., Jansen, J. D., Krause, C., Lewis, S., McGregor, H. V., Mooney, S. D., Moss, P., Nanson, G. C., Purcell, A. & van der Kaars, S. (2013) Palaeoenvironmental change in tropical Australasia over the last 30,000 years—a synthesis by the OZ-INTIMATE group. *Quaternary Science Reviews*, 74, 97–114.

Ripple, W. J., Beschta, R. L., Fortin, J. K. & Robbins, C. T. (2014a) Trophic cascades from wolves to grizzly bears in Yellowstone. *Journal of Animal Ecology*, 83, 223–233.

Ripple, W. J., Estes, J. A., Beschta, R. L., Wilmers, C. C., Ritchie, E. G., Hebblewhite, M., Berger, J., Elmhagen, B., Letnic, M., Nelson, M. P., Schmitz, O. J., Smith, D. W., Wallach, A. D. & Wirsing, A. J. (2014b) Status and ecological effects of the world's largest carnivores. *Science*, 343, 1241484.

Ryan, J., Rashid, A., Torrent, J., Yau, S. K., Ibrikci, H., Sommer, R. & Erenoglu, E. B. (2013) Micronutrient constraints to crop production in the Middle East-West Asia Region: significance, research, and management. *Advances in Agronomy*, 122, 1–84.

Sahanatien, V. & Derocher, A. E. (2012) Monitoring sea ice habitat fragmentation for polar bear conservation. *Animal Conservation*, 15, 397–406.

Saintilan, N., Wilson, N. C., Rogers, K., Rajkaran, A. & Krauss, K. W. (2014) Mangrove expansion and salt marsh decline at mangrove poleward limits. *Global Change Biology*, 20, 147–157.

Saunders, G., Cooke, B., McColl, K., Shine, R. & Peacock, T. (2010) Modern approaches for the biological control of vertebrate pests: an Australian perspective. *Biological Control*, 52, 288–295.

Seehausen, O., Witte, F., Katunzi, E. F., Smits, J. & Bouton, N. (1997) Patterns of the remnant cichlid fauna in southern Lake Victoria. *Conservation Biology*, 11, 890–904.

Selig, E. R. & Bruno, J. F. (2010) A global analysis of the effectiveness of marine protected areas in preventing coral loss. *PLosOne*, 5, e9278.

Selig, E. R., Casey, K. S. & Bruno, J. F. (2012) Temperature-driven coral decline: the role of marine protected areas. *Global Change Biology*, 18, 1561–1570.

Shine, R. (2010) The ecological impact of invasive cane toads (*Bufo marinus*) in Australia. *Quarterly Review of Biology*, 85, 253–291.

Short, J., Kinnear, J. E. & Robley, A. (2002) Surplus killing by introduced predators in Australia—evidence for ineffective anti-predator adaptations in native prey species? *Biological Conservation*, 103, 283–301.

Signorini, S. R., Murtugudde, R. G., McClain, C. R., Christian, J. R., Picaut, J. & Busalacchi, A. J. (1999) Biological and physical signatures in the tropical and subtropical Atlantic. *Journal of Geophysical Research, C. Oceans*, 104, 18367–18382.

Silliman, R. P. & Gutsell, J. S. (1958) Experimental exploitation of fish populations. *Fisheries Bulletin (U.S.)*, 58, 215–252.

Sinclair, A. R. E. (2012) *Serengeti Story: Life and Science in the World's Greatest Wildlife Region*. Oxford University Press, Oxford.

Sodhi, N. S., Lee, T. M., Koh, L. P. & Prawiradilaga, D. M. (2006) Long-term avifaunal impoverishment in an isolated tropical woodlot. *Conservation Biology*, 20, 772–779.

Steen, D. A., Conner, L. M., Smith, L. L., Provencher, L., Hiers, J. K., Pokswinski, S., Helms, B. S. & Guyer, C. (2013) Bird assemblage response to restoration of fire-suppressed longleaf pine sandhills. *Ecological Applications*, 23, 134–147.

Stirling, I. (2011) *Polar Bears: The Natural History of a Threatened Species.* Fitzhenry and Whiteside, Markham, Ontario.

Szabo, J. K., Khwaja, N., Garnett, S. T. & Butchart, S. H. M. (2012) Global patterns and drivers of avian extinctions at the species and subspecies level. *PLoS ONE*, 7, e47080.

Teichman, K. J., Nielsen, S. E. & Roland, J. (2013) Trophic cascades: linking ungulates to shrub-dependent birds and butterflies. *Journal of Animal Ecology*, 82, 1288–1299.

Tingley, R., Phillips, B. L., Letnic, M., Brown, G. P., Shine, R. & Baird, S. J. E. (2013) Identifying optimal barriers to halt the invasion of cane toads *Rhinella marina* in arid Australia. *Journal of Applied Ecology*, 50, 129–137.

Tognetti, P. M., Chaneton, E. J., Omacini, M., Trebino, H. J. & León, R. J. C. (2010) Exotic vs. native plant dominance over 20 years of old-field succession on set-aside farmland in Argentina. *Biological Conservation*, 143, 2494–2503.

Urban, M. C., Phillips, B. L., Skelly, D. K. & Shine, R. (2007) The cane toad's (*Chaunus [Bufo] marinus*) increasing ability to invade Australia is revealed by a dynamically updated range model. *Proceedings of the Royal Society B: Biological Sciences*, 274, 1413–1419.

van Wijk, S. J., Taylor, M. I., Creer, S., Dreyer, C., Rodrigues, F. M., Ramnarine, I. W., van Oosterhout, C. & Carvalho, G. R. (2013) Experimental harvesting of fish populations drives genetically based shifts in body size and maturation. *Frontiers in Ecology and the Environment*, 11, 181–187.

VanDerWal, J., Murphy, H. T., Kutt, A. S., Perkins, G. C., Bateman, B. L., Perry, J. J. & Reside, A. E. (2013) Focus on poleward shifts in species' distribution underestimates the fingerprint of climate change. *Nature Climate Change*, 3, 239–243.

Walker, D. A. (2010) Biofuels—for better or worse? *Annals of Applied Biology*, 156, 319–327.

Walker, L. R. & del Moral, R. (2009) Lessons from primary succession for restoration of severely damaged habitats. *Applied Vegetation Science*, 12, 55–67.

Wallace, A. R. (1876) *The Geographical Distribution of Animals.* Macmillan, London.

Walters, J. R. (1991) Application of ecological principles to the management of endangered species: the case of the red-cockaded woodpecker. *Annual Review of Ecology and Systematics*, 22, 505–523.

Wedin, D. A. & Tilman, D. (1996) Influence of nitrogen loading and species composition of the carbon balance of grasslands. *Science*, 274, 1720–1723.

Whittaker, R. H. (1975) *Communities and Ecosystems*, 2nd ed. Macmillan, New York.

Whittaker, R. J., Bush, M. B. & Richards, K. (1989) Plant recolonization and vegetation succession on the Krakatau Islands, Indonesia. *Ecological Monographs*, 59, 59–123.

Wieder, R. K. & Vitt, D. H. (2006) Boreal Peatland Ecosystems. Springer, New York.

Witte, F., Seehausen, O., Wanink, J., Kishe-Machumu, M., Rensing, M. & Goldschmidt, T. (2013) Cichlid species diversity in naturally and anthropogenically turbid habitats of Lake Victoria, East Africa. *Aquatic Sciences*, 75, 169–183.

Woinarski, J. C. Z., Fisher, A., Armstrong, M., Brennan, K., Griffiths, A. D., Hill, B., Choy, J. L., Milne, D., Stewart, A., Young, S., Ward, S., Winderlich, S. & Ziembicki, M. (2012) Monitoring indicates greater resilience for birds than for mammals in Kakadu National Park, northern Australia. *Wildlife Research*, 39, 397–407.

Woinarski, J. C. Z., Legge, S., Fitzsimons, J. A., Traill, B. J., Burbidge, A. A., Fisher, A., Firth, R. S. C., Gordon, I. J., Griffiths, A. D., Johnson, C. N., McKenzie, N. L., Palmer, C., Radford, I., Rankmore, B., Ritchie, E. G., Ward, S. & Ziembicki, M. (2011) The disappearing mammal fauna of northern Australia: context, cause, and response. *Conservation Letters*, 4, 192–201.

Woinarski, J. C. Z., Risler, J. & Kean, L. (2004) Response of vegetation and vertebrate fauna to 23 years of fire exclusion in a tropical Eucalyptus open forest, Northern Territory, Australia. *Austral Ecology*, 29, 156–176.

World Wildlife Fund. (2014) Living Planet Report 2014. http://wwf.panda.org/about_our_earth/all_publications/living_planet_report/

Young, T. P., Okello, B. D., Kinyua, D. & Palmer, T. M. (1997) KLEE: A long-term multi-species herbivore exclusion experiment in Laikipia, Kenya. *African Journal of Range & Forage Science*, 14, 94–102.

Zhu, X.-G., Long, S. P. & Ort, D. R. (2010) Improving photosynthetic efficiency for greater yield. *Annual Review of Plant Biology*, 61, 235–261.

Zuo, W., Smith, F. A. & Charnov, E. L. (2013) A life-history approach to the Late Pleistocene megafaunal extinction. *American Naturalist*, 182, 524–531.

INDEX

stable states, 79, 82
starfish, 9, 87, 89
starling, 3–4, 7
streptomycin, 105
Sturnus vulgaris, 3, 4
successful transplants, 3, 7
succession, 60, 62, 64
    old-field, 64, 65, 181
    primary, 61, 62, 181
    process of, 55, 56
successional species, 62
sugar cane, 133, 134
sulfur, 123, 139
sulfur cycle, 114, 123, 124
sulfur dioxide emissions, 124
sunlight, 120, 131, 132, 134
surface waters, 76, 137–139
survival, 31, 57, 66, 99, 114, 162, 164, 166, 171

temperature, 9, 21, 23, 63, 74, 80, 95, 111, 127, 130, 139, 141–148
temperature tolerances, 111
tent caterpillars, western, 15–17
terrestrial plant species, 79
territorial behavior, 17–20
thickets, 79–82
Thomson's gazelle, 14
threats, 108, 157, 158, 161
toads, 5
tolerance, 111, 112
top predators, 79, 83–85, 87, 90, 91, 97
toxins, 5, 108, 109
transmission, 35, 106, 108, 109
transplant, 7, 9
transplant experiments, 3, 7, 8
tree rings, 145, 148, 149, 151, 156, 175
trees, 21, 79, 83, 117, 118, 148, 149, 155
    hardwood, 31
    drought-sensitive, 149
    sensitive, 124
trophic levels, 85, 86
trophy hunting, 51

tropical cyclones, 57, 58
tropical oceans, 55, 136, 151
tuberculosis, 105, 106
tundra, 71

United Kingdom, 42, 100, 102, 173
United States, 3, 4, 7, 28, 36, 99, 102, 124, 166, 167
upwelling, 138–139
Ursus arctos, 166

Vancouver Island, 154, 172
vegetation, 23, 60, 63, 92–94, 117, 118, 121, 127, 132, 150, 152, 173, 182
vegetation communities, 90, 153
virologists, 108
virus, 16, 34, 35, 108
volcanos, 54, 60–64, 151

warfare, chemical, 10–11
warfarin, 100
water, 9, 25, 27, 40, 62, 71, 75, 76, 104, 116, 121, 122, 124, 126, 130, 131, 136, 139, 140–144, 156, 157, 165
water column, 75, 76
watersheds, 116, 118
water temperatures, 11, 84
weeds, 28, 37, 64, 101–104
    herbicide-resistant, 102–104, 171
Western Australia, 5, 6, 173
western tent caterpillars, 15–17
West Java, 153, 162
whale populations, 48–50
whales, 21, 48–49, 159, 171
    baleen, 48, 49
    blue, 48
    fin, 48
    killer, 90, 91
whaling, 48–51
wheat, 38, 39, 111, 133, 134, 177
white-tailed deer, 83
wildebeest, 14, 15, 20, 21, 34, 83
wildlife, 37, 51–53, 100
windstorms, 54
winter, 14, 17–19, 37, 73, 87, 161